QUANTUM PHYSICS

FOR BEGINNERS

AN EASY AND COMPREHENSIVE GUIDE TO LEARNING THE FUNDAMENTALS OF QUANTUM PHYSICS. DISCOVER THE Q FIELD THEORY, Q COMPUTING, AND Q MECHANICS

LOEW T. KAUFMANN

© Copyright 2021 BY LOEW T. KAUFMANN
- All rights reserved -

The content contained within this book may not be reproduced, duplicated, or transmitted without direct written per- mission from the author or the publisher. Under no circum- stances will any blame or legal responsibility be held against the publisher, or author, for any damages, reparation, or monetary loss due to the information contained within this book. Either directly or indirectly.

Legal Notice: This book is copyright protected. This book is only for personal use. You cannot amend, distribute, sell, use, quote or paraphrase any part, or the content within this book, without the consent of the author or publisher.

Disclaimer Notice: Please note the information contained within this document is for educational and entertainment purposes only. All effort has been executed to present accurate, up to date, and reliable, complete information. No warranties of any kind are declared or implied. Readers acknowledge that the author is not engaging in the rendering of legal, financial, medical, or professional advice. The content within this book has been derived from various sources. Please consult a licensed professional before attempting any techniques outlined in this book.

By reading this document, the reader agrees that under no circumstances is the author responsible for any losses, direct or indirect, which are incurred because of the use of information contained within this document, including, but not limited to, errors, omissions, or inaccuracies.

TABLE OF CONTENTS

- **INTRODUCTION** 8
- **BEFORE QUANTUM PHYSICS** 14
  - Light and Matter 14
  - Newton's corpuscles of light 14
  - Young's double light experiment 15
  - Maxwell's famous equations 18
  - Electromagnetic spectra 19
  - A nod to thermodynamics 21
- **MAX PLANCK: THE FATHER OF THE QUANTUM THEORY** 24
  - Bohr's atomic model 26
- **THE BIRTH AND FOUNDATIONS OF QUANTUM MECHANICS** 31
  - Understanding the quantum nature of light 31
- **BASIC PRINCIPLES AND LAWS** 37
  - The Huygens-Fresnel principle 39
- **EINSTEIN'S RELATIVITY** 45
  - The equivalence principle 45
  - The problem of inertial forces 47
  - Relativity of Gravity 49
- **THE DISCOVERY OF SCIENTIFICALLY DESTROYED MATERIALISM** 51
- **THE LAW OF ATTRACTION AND** 57
- **QUANTUM PHYSICS** 57
- **THE DOUBLE LIGHT EXPERIMENT** 63
  - A game of probability 66
- **EMISSION OF BLACK BODIES** 69
- **RESISTANCE** 75
  - Stewart-Tolman effect 77

    PIEZOELECTRICITY 78
    LIGHTNING 79

PHOTOELECTRIC EFFECT: 83

EINSTEIN'S THEORY 83

    PHOTOELECTRIC EMISSION CHARACTERISTICS 84
    THE ROLE OF PHOTONS IN PHOTOELECTRIC EMISSION 86
    BONDED SYSTEMS AND BONDING ENERGY 87

QUANTUM REALITY 91

HOLOGRAPHY AND QUANTUM GRAVITY 97

    THE MYTH OF GRAVITY 97

UNIFIED ENERGY AND UNIFIED MATTER 101

    UNIFIED FIELD 101
    UNIFIED MATTER 102
    THE FUTURE OF THE UNIFIED FIELD BASED ON ENERGY STRINGS 102

SUPERCONDUCTORS 105

    NO RESISTANCE 105
    ELECTRIC RESISTANCE SUPERCONDUCTORS 106
    TUNNEL EFFECTS 107
    JOSEPHSON EFFECT 108

EVOLUTION OF ELEMENTARY PARTICLES 111

    THE LAWS OF ENERGY AND MOMENTUM ARE CONSERVED AT THE TIME OF EXIT. 111
    POSTULATES 113
    EVOLUTION OF THE HYDROGEN ATOM. 114

MATHEMATICS, THE LANGUAGE OF PHYSICS 117

    THE ROLE OF MATHEMATICS IN PHYSICS 117
    EXAMPLES OF WHERE APPLIED MATHEMATICS RELATES TO THE WORLD. 118
    THE EXAMPLE OF BERNHARD RIEMANN'S APPLICATION OF MATHEMATICS 120
    THREE OUTSTANDING MATHEMATICIANS: LEONHARD EULER, SRINIVASA RAMANUJAN, AND JOHN VON NEUMANN 121
    LEONHARD EULER, THE MOST PROLIFIC MATHEMATICIAN OF ALL AND "MASTER FIDDLER OF STRINGS." 121
    A CREATIVE MATHEMATICIAN AND GENIUS, SRINIVASA RAMANUJAN (1887 - 1920) 122

JOHN VON NEUMANN (1903 -1957) 123

THE UNIVERSE 125

QUANTUM COMPUTING 131

IS INFORMATION PHYSICAL? 131
WHAT IS A COMPUTER? 131
HOW DO COMPUTERS WORK? 132
HOW SMALL CAN A SINGLE LOGIC GATE BE? 133
CAN WE CREATE COMPUTERS THAT FUNDAMENTALLY USE QUANTUM BEHAVIOUR? 134
WHAT IS A QUBIT? 135
WHAT PHYSICAL PRINCIPLES DIFFERENTIATE CLASSICAL COMPUTERS FROM QUANTUM COMPUTERS? 136

WHAT ARE YOUR QUANTUM THOUGHTS? 139

HISTORY 139
TACTICS OF THE QUANTUM MIND 139
DAVID BOHM 139
PENROSE AND HAMEROFF 140
PENROSE CONTINUES 141
DAVID PEARCE 141
CRITIQUE 141
CONCEPTUAL ISSUES 142
PRACTICAL ISSUES 143
THE PENROSE STATES 143
ETHICAL ISSUES 144

THE QUANTUM DIMENSIONS 147

EXAMPLES AND APPLICATIONS 155

SCHRÖDINGER'S CAT 155
QUANTUM ENTANGLEMENT AND NONLOCALITY 157

TEST REALIZATION OF THE 161

QUANTUM COMPUTER 161

HETEROPOLYMERS 161
ION TRAPS 161
QUANTUM ELECTRODYNAMICS CAVITY 162
NUCLEAR MAGNETIC RESONANCE: 162

QUANTUM DOTS 162
JOSEPHSON JUNCTIONS 163
THE KANE COMPUTER 163
TOPOLOGICAL QUANTUM COMPUTER... 163

THE QUANTUM WORLD OF ROTATING PARTICLES 165

ANGULAR MOMENTUM IN CLASSICAL MECHANICS... 165
SPIN, THE STERN-GERLACH EXPERIMENT AND COMMUTATION RELATIONS 168

FIVE MODERN APPLICATIONS OF QUANTUM PHYSICS 173

BELL'S THEOREM 173
HERE ARE 5 OF THE MOST EXCITING ONES: 174
QUANTUM PHYSICS AS SEEN IN EVERYDAY OBJECTS 177
TOASTER 177
FLUORESCENT LIGHTS.. 178

MODERN PHYSICS 181

MATTER AND ANTIMATTER 181
MEDICAL IMAGIN 185
SOME DIFFERENCES BETWEEN THE RELATED TECHNIQUES: 189
IS BASIC RESEARCH WORTH IT?... 190

MODERN COMPLICATION 193

CONCLUSION 197

QUANTUM PHYSICS

Everything is energy and that's all there is to it.

Match the frequency of the reality you want, and you cannot help but get that reality. It can be no other way.

This is not philosophy. This is physics.

Albert Einstein

INTRODUCTION

There is no doubt that among the various scientists who have approached physics during history, Albert Einstein was able to find solutions, with clear explanations and circumcise, about phenomena unexplained until then. His work literally reshaped the world we knew.

Yet, in the scientific community (and dare we say it, outside of it), Albert Einstein is seen as a kind of demigod - an irrefutable authority that no one dares to touch.

No one except quantum physicists.

If Einstein's theory of relativity is so well regarded and accepted, why do we bother with quantum mechanics? What diabolical reason drives so many contemporary scientific researchers to try to reconcile the worlds of classical and quantum physics?

Well, the only reason why quantum is accepted, but still much discussed, is that it would succeed in solving what classical physics until now could not do and would work to push the boundaries of knowledge and technology beyond the imaginary, into a spectrum that until not long ago we only dared to touch with our thoughts.

September 7, 2014, could have seemed like any other autumn day in the northern hemisphere. But, probably by that time, the leaves were already slightly pale, and the heat of summer was slowly beginning to wear off; the fog of a somewhat more fantastic night vanished, giving way to a perfect autumn day.

Everyone needs to know that September 7, 2014, may have seemed like just another day, but it was actually the day that the Theory of Everything officially came to life! You may have learned about it through a movie about Stephen Hawking's life, or you may have even come across it long before the movie came out.

However, what is essential is that the Theory of Everything is one of the most significant attempts to unify both relativity and quantum theory.

Is what was started in the 1920s by Albert Einstein finally beginning to make sense, after eight decades, under the hands of Stephen Hawking?

The Theory of Everything is, perhaps, one of the most ambitious projects ever. It is one of the theories destined to change every little thing, not only in physics but in science as a whole and, soon enough, in humanity's perception of almost every area of its life.

What the Theory of Everything seeks to do is to finally build a bridge between quantum mechanics and the theory of relativity. Some would go so far as to say that it will "speak the mind of God" (Marshall, 2010) and hold the key for humanity to answer the questions it has been trying to answer for a long, long time now.

There are several candidates for the Theory of Everything. Some of them are implausible to be proven in the equation or in practice, but some of them stand out as sane options that could be the final answer to everything.

Of these, we would like to take the time to name the two most important contenders. We believe it is essential for you to know what the most crucial work in physics is at this time - and as such, we will take the time to expand, just a bit, on these two theories.

One of them is called "String Theory," it says that there is a ten-dimensional space in which we live. This sounds more than astounding, we know, but wait until you hear more.

According to String Theory, pieces of point physics are one-dimensional objects (called "strings"). The theory states that these strings circulate in space and are interconnected. Viewed from a distance, a string behaves like any other ordinary particle (with a mass, charge, etc.) indomitable from the string's vibrational state). For example, one of the vibrational states of the string is signified by gravitation (a particle transmitting gravitational force, in simple terms).

In essence, the Theory of Everything is based on quantum gravity. It aims to address a wide range of questions in fundamental physics - such as what's going on with black holes, how the universe was formed, how to improve nuclear physics, and how to handle condensed matter physics better.

Ideally, string theory will unify gravity and particle physics (one of the significant points that need to be bridged between classical physics and quantum mechanics). At the moment, however, it is not evident how much of this theory can be adapted to the real world and how much it allows for changes in its details.

The other theory that competes with string theory for "Theory of Everything" is Loop Quantum Gravity Theory. This paradigm is heavily based on Einstein's work and was developed in the mid-1980s. To understand it, one must remember that, according to Einstein, gravity is not a force per se but a property of space-time.

Until the Loop Quantum Gravity Theory, there have been several attempts to show that gravity can be treated as a quantum force, like electromagnetism or nuclear energy. However, these attempts have not been successful.

The concept of Loop Quantum Gravity attempts to solve is to ground the intersection of conventional physics and quantum physics on Einstein's geometric formulation. Preferably, this will verify that the universe and period are quantized in the same way that energy and momentum are in the quantum process.

If physicists succeed in proving the Loop Theory of Quantum Gravity, space-time will be depicted with granular space and time, meaning minimal space. In other words, according to the Loop Theory of Quantum Gravity, space is made of a thin fabric of finite loops called "spin networks."

Although string theory seems to be much more prevalent in mainstream media (mainly because some of its proponents are pretty popular, even outside of scientific circles - like Michio Kaku, for example), Loop Quantum Gravity Theory should not be dismissed in any way. On the contrary, most of its implications are related to the birth of the universe, which is also called the Big Bang Theory - and, perhaps, the reason why the TV show of the same name was also named after it.

In addition to string theories and loop quantum gravity, you may also come across other candidates to become the Theory of Everything. Some of them include Causal Dynamical Triangulations Theory, Einstein's Theory of Quantum Gravity, or the Theory of Internal Relativity.

These theories show that active efforts are being made to unify quantum theory and more classical physics, demonstrating that the vast majority of the scientific community pays close attention to quantum mechanics. Who are we to push them away, then? Just because things are still foggy doesn't mean they will stay that way forever. And the whole essence of science, in general, is to dream and aim for something bigger, more complete, and more efficient. That's always been the case and always will be the case.

And when it comes to the ultimate dream, nothing comes so close to the grandeur, supersite, and brevity of the world of quantum mechanics - precisely because it is the one theory that will finally give us a well-deserved push forward in a wide range of discoveries.

What should you believe?

It's up to you. We ask you to learn more about both quantum and classical physics and get your head around it. The beauty of physics and research in this field is that nothing is ever fixed and that theories that might have seemed indestructible have been constantly shattered throughout history, beginning, for example, with the flatness of the Earth.

Believe what you think is true based on your readings and research but stay true to the fallible nature of everything!

CHAPTER 1

Before Quantum Physics

Light and Matter

"The size of an electron stands to a speck of dust as the speck of dust stands to the whole earth" _ Robert Jastrow.

In 1860, James Clerk Maxwell's studies brought about the marriage of electricity and magnetism, revolutionizing our understanding of light.

So, we'll briefly explore the connections between light, colour, and heat. We will encounter a curious 19th-century physical mystery and the first step in falling on the path from classical physics to quantum physics.

Newton's corpuscles of light

Have you ever imagined becoming a celebrity in your job or sport?

Have you ever thought that you could become a champion in both roles?

Sir Isaac Newton was a legend in 2 fields. Not only did he invent the laws of motion, but he also laid the foundation for geometric optics.

Newton believed that light was composed of tiny particles that travel in straight lines called rays. These small corpuscles bounce off mirrors, and when they meet a lens, they bend slightly at the entrance and exit but travel in straight lines along the way.

An alternative theory of light has emerged. Rather than small particles traveling in straight lines through space, this new theory held that light was made up of tiny vibrations in some underlying medium, like the waves of water traveling from the wake of a boat to the shore.

Newton did not accept this wave concept because it seemed inconsistent with his geometric approach of lenses and mirrors.

Given Newton's high reputation among physicists, his rejection of the wave theory of light did not emerge for many decades. After that, however, even he could not pass the experimental tests, and eventually, his classical particle theory of light was definitively discarded.

Young's double light experiment

Now let's go back and explore the light wave theory in detail.

First, we need to establish some basics about wave phenomena.

Imagine you are sitting on a fishing boat on a calm, windless morning.

Suddenly a speedboat whizzes by you, so you notice that your boat starts to sway up and down. This happens because the past speedboat sends a wave through the water.

Now, imagine that another speedboat is going the opposite way at about the same time. You will then notice that when the crests of the two waves arrive simultaneously, your boat floats twice as far. This phenomenon is known as constructive interference.

On the other hand, when the crests of one boat's waves coincide with the depression of the others, your ship doesn't move at all. This phenomenon is called destructive interference.

For all other cases, the height of your boat will be approximately in the middle, as calculated by adding up the measurements of the two individual waves.

This method of arithmetic addition of wave heights is known to physicists as superposition. More generally, the interaction between multiple waves is called wave interference. Furthermore, the unexpected observation of this effect finally allowed the wave theory of light to emerge from Newton's shadow.

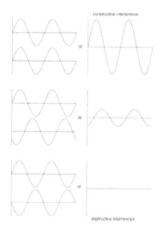

What does this have to do with light?

Well, now imagine that you are in a bedroom with no light. You open the door to a lighted hallway. You would expect a large rectangular shape to glow on the bedroom wall as light passes between the door and its frame.

However, what would happen if you closed the door more so that the space between the door and the frame became thin? In the early 19th century, one of Newton's compatriots, Thomas Young, tried to explain this phenomenon.

He cut a couple of slits in a slightly opaque object and placed it in a dark room. He aimed a beam of light through the slits, then watched what appeared on a screen a few feet away. Instead of seeing two thin stripes appear, a theory predicted by Newton (geometric optics), he noticed a series of strips lined up along the screen like a fence. The explanation for this observation conflicted with the particle theory of light: light behaved like a wave.

Young hypothesized that the initial beam was a wave passing through the room. Interacting with the two slits (points A and B in their original form), each slit served as the source of a new wave, much like the circular pattern formed when water waves pass through a narrow channel. Moving away from the slits, the two waves begin to overlap and interfere. Looking at the darkened screen, the resulting pattern is a series of stripes:

dark when the crest of one wave meets the trough of the other (points C, D, E, and F) and light when the ranges of the two waves coincide.

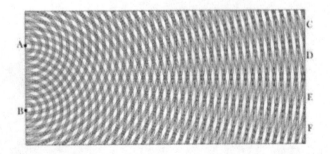

QUANTUM PHYSICS

The image Doctor Young saw on his screen is a diffraction pattern.

Diffraction patterns can be observed whenever two waves interfere with each other, whether it is water waves or light passing through a narrow slit in a dark room.

Although initially treated with skepticism, Young's work gradually gained general approval.

In a short time, he had succeeded in overturning Newton's particle theory of light.

The final blow was delivered about half a century later by Scottish physicist James Clerk Maxwell.

Maxwell's famous equations

By Maxwell's time, physicists had already figured out that static electricity is created whenever you run, say, a piece of amber with a rabbit fur. They had also noticed that the needle of a compass moves whenever a magnet is placed nearby. But, because of the very different nature of these effects, these two phenomena were considered independent and unrelated.

However, some critical observations made it clear that electricity and magnetism could be related around the same time. Maxwell derived a set of four simple equations, which showed that electricity and magnetism were just two sides of the same coin. The two phenomena were always connected by something. This something was called the "electromagnetic field."

Just as a gravitational field allows any mass to pull another, Maxwell's electromagnetic field allows any positive charge to repel positive charges and attract negative ones.

Maxwell showed that a flow of moving electrical charges produces an electromagnetic field that could impact the needle of a compass. He went on to say that if these moving charges increased in speed or changed direction, they would produce an electromagnetic wave that would travel through space.

This wave is a disturbance in the electromagnetic field itself.

Maxwell's classical electrodynamics was very powerful. It explained almost every electrical or magnetic phenomenon known at the time.

For example, it could successfully explain why colours emerged from Newton's prism and why Young's double split formed a diffraction pattern.

Physicists and engineers still use it today to define many electrical and magnetic phenomena with extreme accuracy.

Also, It could be used to calculate the speed at which electromagnetic waves should travel through space.

Maxwell claimed that these waves move at the same speed that physicists had considered for light rays.

After several decades, Maxwell's theory has been confirmed; there was little doubt that light was indeed a wave phenomenon.

Electromagnetic spectra

Today we know that visible light is not the only type of electromagnetic wave out there.

Radio waves picked up by cell phones and microwaves fit a broad electromagnetic spectrum.

The only difference between these different waves is the speed at which they oscillate: this quantity is called frequency and is represented by the symbol f.

According to classical physics, the electromagnetic spectrum is continuous, and any frequency is allowed.

It is also possible to measure the "length" of electromagnetic waves.

The distance to the wave's crests is known as the wavelength (represented by the Greek symbol λ).

When it comes to visible light, red light has the longest wavelength (lowest frequency), while violet has the shortest (highest frequency).

The last quantity needed to describe electromagnetic waves is the speed at which they travel.

The speed of light is indicated by the symbol c.

This is a constant value that never changes.

It is a universal speed limit since nothing can travel faster than c. Mathematically, the three quantities are connected by the equation $c = \lambda f$.

Since all electromagnetic waves travel simultaneously, longer wavelengths have lower frequencies, and shorter wavelengths are higher.

Most light sources, such as the sun, actually emit light that spans a range of frequencies.

Physicists also use unique light sources that emit pure light of only one frequency or monochromatic light.

Compared to the human-sized water waves we observed earlier on the boat, light waves are much shorter.

The wavelength of the orange light emitted by a streetlight is about 60 millionths of a centimetre (0.00006 cm).

It is this "smallness" of light waves that intrigued Newton.

Light waves are very short compared to the size of our mirror, for example.

This means that when light bounces off or passes through it, the deviation from straight-line motion is imperceptible.

Therefore, Newton's geometric optics works well for almost all everyday applications.

It no longer works only when light interacts with microscopic objects, such as Young's thin double slits.

If you travel along the electromagnetic spectrum to gradually longer wavelengths, you arrive at infrared radiation; it is not visible to the naked eye, but instruments such as night vision goggles can easily detect it.

They detect thermal radiation (heat) emitted by objects they see.

We pointed out that Maxwell's classical electrodynamics could explain almost all electromagnetic phenomena we observe every day.

The spectra emitted by heated solids and excited gases, however, are exceptions.

A NOD TO THERMODYNAMICS

Although he will always be remembered as the father of electromagnetism, one of Maxwell's most famous theories, he has had nothing to do with this topic.

In 1873, he addressed the British Association for the Advancement of Science to discuss "molecules." However, he was referring more generally to the concept that gases are composed of small, intensely moving particles.

He claimed that the air in the classroom was filled with molecules traveling in all directions at speeds of about 17 miles per minute. Thus, Maxwell and his contemporaries understood that the temperature and pressure of the air around them were directly proportional to the speed of the gas particles.

There are about 1×10^{23} particles in the volume of a beach ball. Since these velocities will vary somewhat over a specific range, it is more accurate to say that the ambient temperature and pressure are determined by the average speed of all these particles.

The general relationship between particle velocity, temperature, and pressure is called thermodynamics.

It can be classified as the third and final pillar of classical physics; as evidenced by Maxwell's important lecture, its centre is the tiny particles that makeup air.

CHAPTER 2

Max Planck: The father of the quantum theory

ll objects emit electromagnetic radiation, called heat radiation. But we only see them when things are scorching. Because then they also emit visible light. Like glowing iron or our sun. Of course, physicists were looking for a formula that correctly described the emission of electromagnetic radiation. But it wasn't working. Only in 1900, the German physicist Max Planck (1858 - 1947) could draw logical conclusions about it.

The emission of electromagnetic radiation means the emission of energy. According to Maxwell's equations, this release of power must occur continuously. "Continuously" means that any value is possible for the emission of energy. Max Planck assumed that energy emission could only happen in multiples of energy packets, that is, in steps. This led him to the correct formula. To the energy packets, Planck said, "quanta." Therefore, the year 1900 is considered the birth year of quantum theory.

Important: Only the emission (and absorption) of electromagnetic radiation should be in the form of quanta. Planck did not assume that it was composed of quanta because it would have a particle character. However, like all other physicists of his time, he thought that electromagnetic radiation was composed absolutely of waves. Young's double split experiment revealed this, but it was Maxwell's equations that established it.

In 1905 an interloper named Albert Einstein was much bolder. He looked at the photoelectric result confirming that electrons can be thrown out of metals by irradiation with light. According to classical physics, the energy of the eliminated electrons should depend on the intensity of the light. But strangely, this is not the case. The energy of the electrons does not depend on the intensity but on the frequency of the light. Einstein could explain this. That's why we go back to Max Planck's quanta again. The energy of each quantum depends on the frequency of electromagnetic radiation. The higher is the frequency; the higher is the energy of the quantum. Einstein assumes now, in contrast with Planck, that electromagnetic radiation itself is made of quanta. The interaction of a single quantum with a single electron on the metal surface causes this electron to be knocked out. The quantum releases its energy to the electron. Thus, the energy of the knocked-out electrons depends on the frequency of the incident light. The higher the frequency, the higher the energy of the quantum. Einstein now assumes, in contrast to Planck, that electromagnetic radiation itself is made up of quanta. The interaction of a single quantum with a single electron on the metal surface causes this electron to be knocked out of the way. The quantum releases its energy to the electron. Thus, the energy of the knocked-out electron depends on the frequency of the incident light.

However, skepticism was great at first. Because electromagnetic radiation would then have both a wave and particle character, but another experiment also demonstrated its particle character. This experiment was conducted with X-rays and electrons by American physicist Arthur Compton (1892 - 1962) in 1923. As mentioned above, X-rays are also electromagnetic radiation, but they have a much higher frequency than visible light. Therefore, X-ray quanta are very energetic. That is why they can colonize a particular shape. But that makes them so threatening. Compton was very clever in seeing that X-rays and electrons behave similarly to billiard balls when they meet. This again showed the bit character of electromagnetic radiation. Thus, their dual nature, the supposed "wave-particle dualism," was finally recognized. By the way, Compton introduced the term "photons" for quanta of electromagnetic radiation.

What are photons? This is still unclear today. Under no circumstances should they be imagined as small spheres moving forward at the speed of light. Because photons are not found in space, so they are never in a specific place. Here is a quote from Albert Einstein. Although it dates from 1951, it also applies to today's situation: "Fifty years of hard work have not brought me any closer to answering the following question:

"'What are light quanta?
Today, every Tom, Dick, and Harry imagines they know. But they are wrong."

Bohr's atomic model

Nowadays, we take atoms for granted, but their existence was still controversial until the beginning of the 20th century.

Except for the ancient Greeks who were already talking about atoms in the 5th century BC, especially Leukipp and his student Democritus. They thought that matter was composed of small, indivisible units. They called these atoms (in ancient Greek, "átomos" = inseparable). In his miraculous year 1905, Albert Einstein not only presented the particular theory of relativity and solved the mystery of the photoelectric effect,

but he was also able to explain Brownian motion.

In 1827, Scottish botanist and physician Robert Brown (1773 - 1858) discovered that dust particles visible only under a microscope make jerky movements in the water. Einstein was able to explain this because much smaller particles, which are not even visible under the microscope, collide in large numbers with dust particles. This is subject to random fluctuations. The latter leads to jerky movements. Therefore, the invisible particles must be molecules. The explanation of Brownian motion was considered their validation and, therefore, the validation of atoms.

In 1897, British physicist Joseph John Thomson (1856 - 1940) discovered electrons as atoms and developed the first atomic model, the so-called raisin cake model. Thus, atoms consist of a uniformly distributed positively charged mass in which negatively charged electrons are embedded like raisins in a cake. This was falsified in 1910 by New Zealand physicist Ernest Rutherford (1871 - 1937). With his experiments at the University of Manchester, he was able to show that atoms are nearly empty. They comprise a small positively charged nucleus. Around it is the electrons. They should rotate near the nucleus as planets rotate near the sun. Any other form of motion was inconceivable at that time. This brought physics into a deep crisis. Because electrons have an electric charge, and a circular motion leads them to release energy in electromagnetic radiation. Therefore, electrons should fall in the nucleus. Hence the deep crisis because there should be no atom.

In 1913 a young colleague of Ernest Rutherford, Danish physicist Niels Bohr (1885 - 1962), tried to explain the stability of atoms. He transferred the idea of quanta to the orbits of electrons in atoms. This means that there are no random orbits around the nucleus for electrons, but only certain orbits are allowed. Each one has a specific energy. Bohr assumed that these allowed orbits were stable because the electrons on them do not emit electromagnetic radiation. Without, however, being able to explain why this should be the case.

Nevertheless, his atomic model was initially quite successful because it could explain the so-called Balmer formula. It has long been identified that atoms absorb light only at certain frequencies. These are called spectral lines. In 1885, the Swiss mathematician and physicist Johann Jakob Balmer (1825 - 1898) found a formula by which the frequencies of spectral lines could be described correctly. But he could not explain them. Bohr then succeeded with his atomic model, at least for the hydrogen atom. This is because photons can be excited by other photons, making them jump to higher energy orbits. This is the famous quantum jump, the minor possible jump ever. Since only certain orbits are allowed in Bohr's atomic model, the energy and frequency of the exciting photons must correspond precisely to the energy difference between the initial and the excited orbit. This explained Balmer's formula. But Bohr's atomic model quickly reached its limits because it only worked for the hydrogen atom. The German physicist Arnold Sommerfeld (1868 - 1951) extended it. However, it still represented an unconvincing mixture of classical physics and quantum aspects. Moreover, he still could not explain why certain electron orbits had to be stable.

Sommerfeld had a young assistant, Werner Heisenberg (1901 - 1976), who, in his doctoral thesis, took up the Bohr atom model extended by Sommerfeld. Naturally, he wanted to improve it. In 1924 Heisenberg became an assistant to Max Born (1882 - 1970) in Göttingen. The breakthrough came a short time later, in 1925, on the island of Heligoland, where he cured his hay fever. He explained the frequencies of spectral lines, including their intensities, using the so-called matrices. He published his theory in 1925 and its leader Max Born and Pascual Jordan (1902 - 1980). This is considered the first quantum theory and is called matrix mechanics. I will not explain it in more detail because it is not very clear and also because there is an alternative mathematically equivalent to it, much more precise and understandable. I am talking about wave mechanics developed in 1926, only one year after matrix mechanics, by Austrian physicist Erwin Schrödinger (1887 - 1961).

The Schrödinger equation

Before getting to Erwin Schrödinger's wave mechanics, it would make more sense to mention the work done by French physicist Louis de Broglie (1892 - 1987). In his doctoral dissertation, which he completed in 1924, he made a bold proposal. As explained in the penultimate area, the wave-particle dualism was a characteristic exclusively of electromagnetic radiation. Why, therefore de Broglie, should not be valid also about the matter? Why, therefore, should matter not have a wave character besides its real particle character? The examining committee of the famous Sorbonne University in Paris was not sure to approve it and asked Einstein. Nevertheless, he was deeply impressed, so much so that de Broglie was awarded a doctorate. Unfortunately, however, he could not present any elaborate theory for matter waves.

Erwin Schrödinger then succeeded. In 1926 he introduced the equation that bears his name. The circumstances surrounding his discovery are unusual. Schrödinger is said to have discovered it in late 1925 in Arosa, where he was with his lover.

Schrödinger's equation is at the heart of wave mechanics. As mentioned above, it is mathematically equivalent to Heisenberg's matrix mechanics. But it is preferred because it is much easier to use. There is a third, more abstract version developed by the English physicist Paul Dirac. All three versions together form the non-relativistic quantum theory called quantum mechanics. As you rightly suspect, there is also a relativistic version.

The Schrödinger equation is not an ordinary wave equation since it is used, for example, to describe water or sound waves. But mathematically, it is very similar to a "real" wave equation. Schrödinger could not explain why it is not identical. He had developed it more by intuition. According to the motto: How could it be a wave equation for electrons? This can also be called creativity. Very often, in the history of quantum theory, there was no rigorous derivation. Instead, it was a matter of trial and error until the equations that produced the desired result were found. Oddly enough, a theory of such precision could emerge from this. However, as I will explain in detail, the theory is also dominated by problems that have not been solved.

The solutions of the Schrödinger equation are the so-called wave functions. Only with them has it been possible to explain the stability of atoms

QUANTUM PHYSICS

convincingly. Let's consider the simplest atom, the hydrogen atom. It is composed of a proton as the nucleus and an electron moving around it.

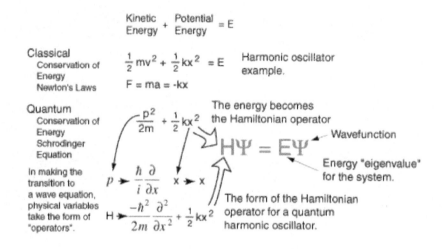

CHAPTER 3

The Birth and Foundations of Quantum Mechanics

Understanding the Quantum Nature of Light

Physics seeks to interpret the laws that affect motion and matter. However, quantum physics is trying to understand the behaviour of smaller particles and how they move. Such particles contain things like electrons, protons, and neutrons.

A. Quantum Physics in the Smallest Details

In its emphasis on microscopic particles, quantum physics explains the particles that make up tiny particles. The rules governing macroscopic structures have been wrong to set precedents for smaller domains since the 20th century. The word "quantum" originates from the Latin term meaning "how much." It is used in physics to refer to the tiny units of matter and energy whose action is predicted and observed in quantum physics.

Conditions that exist firmly and constantly, for example, space and time, also have values. However, they appear to be of the slightest degree.

The quantum model of the atom is even more complicated than we have seen before; instead of orbiting the nucleus like stars, electrons orbit in fuzzy, less defined, or cloud-like formations. In addition, the final configurations we learned from the electron sequence (citing the number of electrons in the outer shell) are generally more like probabilities than tight, rigid formations.

We bring this up when we address the quantum nature of light to define the term quantum physics, so you can understand that its purpose is to show the numerical probability of the electron's place at any available time. Therefore, when the word is combined with "the essence of light," you should have a strong understanding of the general operating principle.

B. Singular to quantum physics

Considering anything that can plausibly influence the physical cycles that happen is unique to quantum material science. For example, in what is perceived as wave-molecule duality, light waves make indistinguishable particles, and these sections additionally show up as waves. Said otherwise, light has the two qualities of particles and waves, and both clarifications can describe the activity of light.

In quantum tunnelling, the matter can start with one area then onto the next without traveling through mutual space. This offers a pathway for advanced application where data can immediately cross large separations. We find that much of the universe can be spoken of as a progression of probabilities through quantum material science.

There is a wide range of fields of quantum materials science. The one that shines, especially on the conduct of light (photons), is Quantum Optics. By investigating Quantum Optics, you will discover that examples of the development of individual photons (light trees) legitimately influence hot light. The severe and flexible tool known as LASER is just one of many side effects.

This contrasts with the most common study of light, Classical Optics, acquired by Sir Isaac Newton, where light was detected as having solitary particle belongings, which means it travelled in a straight line, returned from objects it emanated in contact with and spread through things with minimal resistance.

C. Photons

To understand more easily what is being suggested when using the term photon, let's direct our focus to the Photon Theory of Light. A photon is a vigilant group (or quantum) of electromagnetic vitality (or light) in this specific sense.

Existing in a vacuum and in constant motion, photons have a constant speed of light for all eyewitnesses. Therefore, it occurs at the speed of light in a vacuum (most often referred to as the speed of light), which is used.

$C = 2.998 \times 10^8$ m/s long.

Based on the photon theory of light, the main attributes of photons are as follows:

- They are propelled at a constant velocity, $c = 2.9979 \times 10^8$ m/s (speed of light) in free space.

- They are known to have zero mass and zero resting vitality.

- They transmit vitality and energy, which relate to the recurrence nu and frequency lambda (and p, strength) of the electromagnetic wave from $E = h\nu$ for and $p = h/\lambda$

- They can be destroyed or created when radiation is absorbed or emitted.

- They can have particle-like interactions, e.g., electron impacts and other moments.

D. Quantum Optics. Basic understanding

To better understand the quantum properties of light, it may be helpful to apply some of the relative cycles (retention, emanation, and invigorated outflow) to lasers since this is one of the most notable uses of quantum optics. In general, these three equivalent attributes could be summarized to other light sources in varying degrees.

Electronic advances are usually the types of advances that transmit or ingest remarkable light. Just imagine an electron moving between measured nuclear vitality levels to perceive how it works.

For the laser to function productively, the invigorated light stream is significant. The emanation of animated light is utilized to give the improvement needed to perform the imaging job appropriately.

The exceptional property known as cognition is the consequence of the animated emanation measure. Typical updating triggers emanation occasions that are accountable for giving the improved light. These partners are the photons released in the ideal advancement arrangement, where every photon has the last stage ratio.

This kind of intelligibility (relative arrangement) is characterized in two different terms: world cognition and spatial solidity. Both end up being highly critical in the obstruction advancement used to produce 3D images.

Note: Ordinary light is not intelligent because it starts from autonomous iotas that transmit about 10-8 seconds in time scales. While there might be some level of rationality in the sources chosen, for example, the green line of mercury and the sprinkling of other valuable phantom sources, their consistency is not about what is contained in the laser.

A few specific features that are unique to laser light include:

1. **Cognition**: This is the property in which different parts of the laser tree are identified in a phase relationship. To the point where, if maintained for a long enough period, impedance impacts can be seen or recorded photographically. Intelligence is the factor that makes the possibility of visualizations conceivable.

2. **Monochromatic**: a frequency, laser light starts from the invigorated outflow of a solitary scope of nuclear vitality levels.

3. **Collimated**: Since they must pass through mirrors a couple of times at surprisingly opposite edges, intensification-influenced modes are known for their capacity to bounce back between the reflected finishes of the laser hole firmly. For this reason, laser beams have been thought to be extremely tight and limited in their ability to expand.

E. **Photon and Probability:** There are two ways in which probability can be applied to photon activity: probability can quantify the conceivable number of photons in a specific state or can be used to estimate the possibility of a single photon being within a particular state.

Since the previous understanding opposes Newton's principle of conservation of energy, the last translation is the other more practical option.

Following physicist Thomas Young's creation during the 1800s with a double-edged investigation, Paul Dirac (1902-1984), a hypothetical British physicist and one of the key pioneers of quantum material science, speaks of this norm in its updated variant.

Even before quantum material science was found, individuals realized that a connection between light waves and photons had to have a real character. Nevertheless, it was not so sure that wave work provided data on the probability of a photon in each situation rather than the absolute plausible number of photons in that area.

This is an important distinction and can be clarified as follows. Suppose we have a light emission consisting of numerous photons isolated in two pieces of equivalent strength. Half the probable number of photons should be appropriate for each distribution if the bar is connected to a reasonable measure of photons. As the parts are made to contact each other, a photon in one segment should disrupt the other.

According to the old theory, the two photons should cancel each other out or generate four photons. Thus, any result would contradict the principle of conservation of energy.

Thus, under the new theory, since the photon only slightly affects both, the question of relating the wave function to the probabilities for a single photon becomes a non-problem. Each photon can only cause interference in this system, preventing the occurrence of two-photon potentials.

CHAPTER 4

Basic Principles and Laws

s any excellent reading on electromagnetism will tell you, a single slit, or pinhole, will produce an interference pattern. However, one is less pronounced than that provided by a larger slit. This is a natural experiment you can do yourself:

- Take a piece of paper.
- Make a small round hole with a pin or needle.
- Look toward a light source.

You will observe the image of the light source surrounded by several concentric coloured bangs (the colours only appear because, fortunately, the world we live in is not monochromatic).

In this case, you see that even though we are only dealing with a single slit, you can notice weak but still clearly visible secondary minimum and maximum bangs.

An important point to keep in mind is to avoid a common misconception (which is frequently promulgated in some popular science readings), which states that only two or more slits can produce interference bangs, whereas, for the single slit, the interference effects disappear. This is not entirely correct.

True, it is easier to produce more pronounced interference patterns with more than one slit (or pinhole). However, for most applications, especially when the wavelength of the incident wave is much smaller than the size of the aperture, these effects can be neglected.

However, strictly speaking, a single slit also produces small diffraction and interference phenomena.

An elegant explanation of how interference arises, even for a single slit, goes back to the French physicist A. J. Fresnel. He borrowed an idea of Huygens (hence the name "Huygens Fresnel principle"), according to which every single point on a wavefront should be considered itself a point-source of a spherical wave.

Along with the aperture, they emit their spherical wavefronts simultaneously, but when viewed from one position on the screen, they add up to produce an interference pattern. The reason for this is not so difficult to visualize.

Since all the fonts are initiated at different positions along with the aperture, they will also travel an extra path length, implying that they have various phase shifts when they overlap on the screen.

For example, where we saw the two source paths from the edges. Fresnel showed that if you add up all the spherical wavefronts coming from the aperture points of the single slit and project them onto all the points together on the detector screen, you get the well-known diffraction and interference patterns.

The Huygens-Fresnel principle

If we recapitulate the same experiment with a slit size close to the wavelength of our incoming wavefront, we see that the interference bangs disappear.

Only when the slit size is equal to or less than the wavelength is the bangs absent. The slit is so low that only a single point source can form a spherical wavefront with a wavelength equal to the slit size, and there can be no path difference and phase shift with some other source that could produce the interference pattern. However, the diffraction has instead become very large so that the photons will move over a relatively large area on the detector screen, according to a bell-shaped distribution called the diffraction envelope.

The parameters that determine the angular dependence of the interference pattern are: first, the aperture size versus wavelength (here: a=3); second, the spacing d between the slits (here: d=3 a); and, of course, the number of slits. The three curves represent diffraction cases at 1, 2, and 10 slits, respectively. The intensities were normalized to all unit cases.

For the 1-slit case, you see that there are some weak but discernible secondary side peaks. They reduce almost to the diffraction envelope.

For the two slits, as in the case of Young's double slit experiment, we get more pronounced bangs. You can see how the one-slit model 'envelopes' the two-slit model. However, note that it would not be correct to say, as we often hear, that when we switch from the double-slit case to the single-slit case, the interference phenomena disappear. This is not, in general, the case. What happens is that we go back to the single-slit case, which contains far fewer bangs, but could still have other interference bangs as well (and in this case, it does). Again, interference is not a phenomenon specific to the double (or multiple) slit experiments.

Interference does not disappear if one slit is covered; it simply becomes weaker than it is with multiple slits.

Finally, in the case of 10 slits, the curve of the two slits turns out to be the envelope of the angle of the ten slits. Thus, it can be seen that this is a trend and a more general phenomenon that results from the interaction between diffraction and interference. In general, N-slits bangs and their spacing arise because of this combined effect between diffraction and interference.

These were just a few examples to outline, at least intuitively, how wave interference works.

We might ask: What happens with a particle if we want to know its precise position in space? For example, we determine the exact position by passing it through a small pinhole, as Isaac Newton did with photons in his investigations into the nature of light. If a particle passes through that single small hole on a piece of paper, we are entitled to say that we can determine its precise position in space. Because on the other hand, we identify that we cannot forget the wave aspect of the particle because of the wave-particle duality. When a particle, even a material particle, passes through this pinhole, it will be equally diffracted. Later, it will be positioned on the detector screen according to an interference pattern.

If instead of dealing with slits, we take a tiny round hole of comparable size to some multiple of the wavelength, we get circular interference bangs. This is an inherent and inevitable effect for all types of waves.

The pinhole, as a detector of a particle's position, cannot avoid interference.

According to de Broglie relation, suppose particles must connect to a wave, conceiving it as a wave packet. In that case, we will always have interference, even with only one hole or slit and even with only one particle.

Compare the two cases where the pinhole determines the particle's position with an aperture of $a=2\lambda$ (high precision) and $a=20\lambda$ (low accuracy), with, as usual, the photon's wavelength or, in the case of a matter wave, the de Broglie wavelength.

If the pinhole is small, while it will decide the situation of a molecule more precisely, it will likewise deliver a generally expansive diffraction pattern that provides the area of the photon on the unsafe screen. Thus, we can usually know where the photon or matter molecule experienced the piece of paper inside the space enclosed by the pinhole gap. However, it will still be uprooted horizontally on the screen due to diffraction and the wonders of impedance.

Of course, no interference pattern is visible with a single particle producing a single dot on the screen. However, as we learned with single-photon diffraction at the double slit, the probability of finding this spot is in one-to-one correspondence with the intensity of the interference bangs that many particles produce.

Recall also that we cannot predict where precisely this spot will appear.

On the other hand, if the pinhole is large, the bangs will become less pronounced. This is because we will know where the photon will hit the screen with relatively good accuracy, meaning that it only "felt" a slight shift along with the screen.

However, in doing so, we will lose our ability to determine where precisely the particle passed through the pinhole since it is no longer a pinhole but a large hole. Thus, there is no way, ever, even in principle, to get the precise measurement of the particle's position and, at the same time, avoid the production of interference bangs (or interference circles, as in the case of a circular aperture or significant diffraction effects). As a result, you will always get a more or less pronounced bell-shaped or peaked distribution of white dots on the screen. This is not because we do not have a sufficiently accurate measuring device but because it is a consequence of the intrinsic wave nature of particles. It is a universal law of nature that it is impossible to believe we can pass a wave through a slit and not observe interference and diffraction phenomena.

Heisenberg's uncertainty principle was explained using the single slit (or pinhole) diffraction experiment.

Now, this could also be interpreted as follows: The particle, once it passes through the slit, will acquire extra momentum, λp, along the vertical axis. It does not occur due to the interaction of an external force or, as we might naively imagine, due to an effect of communication, deflection, or bounce of the particle with the edges of the slit because, in that case, we would observe a random distribution but not an interference pattern. This extra λp moment, which moves the particle along with the detection screen, is due solely and exclusively to the wave nature of matter and light. We could also interpret this as 'scattering' of the particle, but we should remember that this is misleading terminology, since there is no scattering force. No power scattering interaction from outside is necessary for this to happen.

Where does this extra momentum λp come from? It is simply the uncertainty we have about the momentum of the particle in the first place. It is inherent uncertainty in the properties of any particle because of their wave nature. It is the only conclusion possible to avoid violating the principles of conservation of momentum and energy.

The point is that we will never be able to determine with extreme precision.

- That is, with an infinitely small slit of size $\lambda x=0$ - without blurring the momentum because, by doing so, we will inevitably spread the plane wavefront, whose wavelength is given by de Broglie relation.

We will inevitably move it according to a statistical law that reflects the diffraction and interference laws.

Thus, we must conclude that the trimmer is uncertain when determining the particle's position (the size λx of the slit). The more significant the diffraction effects are, therefore, the more meaningful the change in momentum. (On the other hand, if I want to know the momentum of the particle with a small uncertainty (λp small), we will have to open the aperture of the slit (λx large) to reduce diffraction. However, we will never be able to accurately determine both the momentum and position of a particle simultaneously. We must choose whether we want to focus on one or the other; we can never get both. Again, this is not because we are perturbing the system but because we are dealing with waves.

CHAPTER 5

Einstein's Relativity

In 1907, only two years after developing the theory of special relativity, Einstein had the idea that he would describe as "the happiest of his entire life." In this inner vision, there appeared to him what would prove to be the essential physical basis of general relativity, even though it would take him nearly ten years to mathematically develop the theory. Einstein realized that:

"If a man falls freely, he does not feel his weight."

Even the expression "free fall" is eloquent. Although one is permanently attached to a gravitational field, attracted to the Earth from the point of view of Newtonian theory, one finds freedom when one falls. It is this freedom that those who pursue freefall as a hobby seek to see and feel, even if it is only partly due to air resistance. It is, of course, the astronauts in "weightlessness" who truly experience for a long time this feeling of no longer having weight, of no longer being subject to the pull of the Earth. However, Einstein's big idea was the understanding that if we jump up during the moment of our jump, we experience this "weightlessness." In other words, there is no difference in principle between a ship orbiting the Earth and a ball we throw here on Earth: both are in free fall; both are, for the duration of their motion, satellites of the Earth.

The equivalence principle

Understanding this universal phenomenon led Einstein to formulate the principle of equivalence, according to which a gravitational field is locally equivalent to an acceleration field. To obtain this principle, he drew on a fundamental property of gravitational fields already highlighted by Galileo and included in Newton's equations: the acceleration communicated to a body by a gravitational field is independent of its mass.

After the development of special relativity, the need to generalize the theory seemed inevitable for multiple reasons. First, relativistic unification was far from complete. If free particle mechanics and electrodynamics finally satisfied the same laws, this was not the case for Newton's theory of universal gravitation, otherwise the centrepiece of classical physics. Newton's equations are invariant under Galileo's classical transformation but not under Lorentz's. Thus, physics remained split in two, in contradiction to the principle of relativity, which requires the validity of the same fundamental laws in all situations.

Moreover, the Newtonian theory is based on some assumptions in contradiction with the principle of relativity: it is so for the concept of Newtonian force, which acts at a distance propagating instantaneously at an infinite speed. Thus, constructing a relativistic theory of gravitation seemed a logical necessity to Einstein (and other physicists).

Another problem was just as profound: the relativistic approach explicitly poses the problem of changes in reference systems and their influence on physical laws. But the answer provided by special relativity is only partial. It considers only reference frames in uniform translation, at constant velocity concerning each other. However, the natural world constantly shows us rotations and accelerations, from multiple forces at work (such as gravity), or inversely, causing new forces (such as the forces of inertia).

What are the transformation laws in the case of accelerated reference frames? Why would such reference frames not be as good at writing the laws of physics as inertial reference frames? The answer is that this question requires a generalization of special relativity.

The originality of Einstein's approach had been, in particular, to bring together two problems: constructing a relativistic theory of gravitation and generalizing relativity to non-inertial systems in a single effort. The principle of equivalence has made this unity of approach possible: if the acceleration field and the gravitational field are locally indistinguishable, then the two problems of describing changes in coordinate systems, including those accelerated and those subject to a gravitational field, are reduced to a single issue. But such an approach is not reducible to "making Newtonian gravitation relativistic." While some physicists might have hoped, at the time, that the problem of Newton's theory could be solved by a simple reformulation, introducing a force that propagates at the speed of light, it is the whole picture of classical physics that Einstein set out to reconstruct with general relativity. Better still, it was a new kind of theory that he developed for the first time: a theory of a framework (curved space-time, now a dynamical variable) concerning its content, and no longer just an approach of "objects" in a pre-existing rigid framework (as Newton's absolute space was).

Why such a radical choice? Undoubtedly because special relativity itself was unsatisfactory at least on one essential point: the space-time that characterizes it, even if it includes in its description space and time no longer absolute taken individually, remains absolute if taken as a four-dimensional "object." However, inspired in particular by the ideas of Ernst Mach, Einstein had come to think that an absolute space-time could not have a physical meaning, but rather its geometry should be in correspondence with its material and energy content. Thus, a reflection on the problem of inertial forces, which had led Newton to introduce absolute space, led Einstein to the opposite conclusion.

THE PROBLEM OF INERTIAL FORCES

The existence of inertial forces intensely poses the problem of the absolute or relative nature of motion and, ultimately, space-time. Mach's ideas in this area had a profound influence on Einstein. For Mach, the relativity of motion did not apply only to uniform motion in translation; instead, all movement of any kind was by essence relative (Poincaré and, long before him, Huygens had arrived at the same conclusions).

This proposition may seem to contradict the facts. Suppose it is clear, since Galileo, that it is impossible to characterize the state of the inertial motion of a body in an absolute way (only the velocity of a body to another has a physical meaning). In that case, it seems different in the case of accelerated motions. Thus, when considering a body spinning on itself, the existence of its rotational motion appears to be able to be perceived intrinsically to the body. There is no need for another reference body: it is enough to check whether a centrifugal force that tends to deform the rotating body appears.

Reconsidering the mental experiment of Galileo's ship, the difference between inertial motion and rotational motion is accentuated. No investigation conducted in the cabin of a boat traveling in uniform and rectilinear motion concerning the Earth can determine the existence of the vessel's motion. As Galileo had understood, "motion is like nothing." Relative motion can only be determined by opening a porthole in the cabin and observing the passage of the Earth. But now, if the boat accelerates or turns on itself, all objects in the cabin will be pushed towards the walls. Thus, the experimenter will know that there is motion without having to look outside. Therefore, accelerated motion seems definable by a purely local experiment.

This argument led Newton to allow an unlimited space to be defined, in opposition to Leibniz (later Mach), for whom defining a space independently of the objects it contains could not make sense.

Mach proposed a solution to the problem utterly different from Newton's. Starting from the principle of relativity of all motions, he came to the natural conclusion that the spinning body, inside which inertial forces appear, must not spin concerning a particular absolute space but with respect to other material bodies. Which ones? They cannot be narrow bodies whose distribution fluctuations would cause visual fluctuations of inertial systems. This is unacceptable because it is easy to check the coherence of these systems over large distances. Thus, if we look, motionless relative to Earth, at the night sky, we do not see the stars turning.

However, if we turn around on ourselves, we feel our arms expanding due to the inertial forces, and as we look up at the sky, we see it turning. This was Mach's initial observation: within the same frame of reference, the arms rise, and the sky turns, and this will be true for two points on Earth separated by thousands of miles. Mach suggested, then, that the standard frame of reference is determined by the whole of the extraneous matter of the bodies "at infinity," of which the cumulative gravitational influence would be at the origin of the inertial forces. In other words, the body would rotate about a reference frame, not absolute but universal. An absolute motion would be defined, independent of all objects. However, Mach claimed that every move is relative, remaining defined for an "object," even if it is the universe in its totality.

Einstein's proposed solution, that of the equivalence principle and general relativity, incorporate some of these ideas while ultimately distancing itself from Mach's principle, even though its premises were identical. First, the input of matter and energy throughout the universe determines the geometric structure of space-time. Then the motions of bodies are realized within the framework of this matter-related geometry.

Relativity of Gravity

Let us now return to Einstein's great idea of 1907. If an observer goes down in free fall within a gravitational field, he no longer feels his weight; that is, he no longer feels the existence of this field itself. This observation, which today may seem evident to us - we have all seen, on television or in movies, weightless astronauts floating in their spacecraft, and the objects they drop moving away from them at a constant speed - is nevertheless revolutionary, because it implies that gravity does not exist, that its very existence depends on the choice of a frame of reference.

It thus distances itself from the previous concept of gravity. What is more absolute than a gravitational field in the Newtonian model? Newton recognized gravity as universal; here is a physical phenomenon whose existence does not seem to depend on such an observation condition.

However, if we freely drop an enclosed area within a gravitational field, and then put a body in motion at a certain velocity toward this area, the body will move in a straight line at a constant rate for the walls of the enclosure; an initially immobile body (again, toward the walls) will remain so during the motion of dropping the compartment. In other words, all the experiments we can do would confirm that we are in an inertial reference frame! Thus, however universal, a reasonable choice of coordinate system can only cancel out gravity. What Einstein realized in 1907 was that even the existence of gravity was relative to the will of the coordinate system.

CHAPTER 6

The discovery of scientifically destroyed materialism

The quantum model of the universe is an attempt to rid the enormous explosion of its creationist suggestions. Proponents of this model depend on perceptions of quantum material science (subatomic material science). In quantum material science, one can very well see that subatomic particles appear and disappear precipitously in a vacuum. If some physicists decipher this perception so that matter can emerge at the quantum level, it is a property that hints at the problem. Some physicists attempt to clarify points from non-presence during the production of the universe as a property identified with an issue and to talk about it as a characteristic of the customary laws.

However, this logic is impossible and cannot clarify how the universe was conceived. William Lane Craig, creator of The Big Bang: Theism and Atheism, makes it clear why:

A quantum-age material of the mechanical vacuum is a long way from the standard thought of a 'vacuum,' equal to nothingness. This is 'nothingness,' and material particles do not emerge from the void along these lines.

Matter does not exist in early quantum material science. What has happened is that ecological vitality suddenly becomes matter, and afterward, as if from nothing, it becomes vitality again. Thus, there is no such thing as an existential condition from nothing, as is guaranteed.

As indicated by Isaac Newton, the light was the development of a substance known as a body. The reason for ordinary Newtonian material science - recognized until the popularization of quantum physical science - was that light included a grouping of particles. In any case, James Clerk Maxwell, a nineteenth-century physicist, recommended that light exhibited a wave-like development. Quantum speculation obliged this more basic conversation in material science.

In 1905, Albert Einstein assumed that light had quanta or little bundles of essentialness. These bundles of imperativeness were called photons. But represented as particles, photons were believed to continue in the wave development proposed by Maxwell in 1860. Consequently, the light was wandering progress between waves and atoms (George Gilder), an articulation that showed a significant irregularity in the measure of Newtonian material science.

After Einstein, German physicist Max Planck dissected light and stunned the whole coherent world by discovering it was both a wave and an atom. But, as demonstrated by this thought, which he proposed under the name of quantum theory, vitality was conveyed as a deterrent and discrete package instead of straight and reliable.

In a quantum event, light exhibited both particle and wave properties. For example, a wave in space went with the particle known as a photon. However, light traversing space as a wave went around as a functioning particle when it encountered a tangle. To put it bluntly, it presented itself as essential until it encountered an obstacle. At that point, it presented itself as particles as if they contained tiny material bodies reminiscent of grains of sand.

Amit Goswami says this about the discovery about the nature of light:

Just when light is seen as a wave, it seems to be ready to end up in (in any case two) positions simultaneously, as if it had experienced the opening of an umbrella, making a diffraction plane. In any case, when we get it on photographic film, it shows sagaciously point by point as a rod of atoms. Consequently, the light should be both a wave and an atom. This is one of the limitations of ancient material science and is far from language. Reference is also made to objectivity; does light or what light depends on how we see it?

Specialists, at this point, did not recognize that the question was included in inorganic and self-absorbed particles. Thus, quantum physical science had no materialistic note because there were useless things.

De Broglie's revelation was remarkable; in his investigation, he saw that even subatomic particles exhibited wave-like properties. Even particles like the electron and proton had frequencies. Figuratively speaking, despite the materialist belief, there were unimportant overhangs of essentiality within the atom, which authenticity called fundamental matter. Equivalent to light, these tiny particles in the particle behaved like waves during a part of the time and showed the properties of particles in others. Thus, contrary to typical desires, the fundamental problem of the molecule could be recognized in explicit events.

This essential exposition showed that what we imagine as this current truth was a shadow. The problem had moved away from the realm of material science and into the otherworldly.

Physicist Richard Feynman portrayed this fascinating reality about subatomic particles and light: "Little by little, we are discovering how electrons and light act. However, how could I have the ability to name them? When I state that they behave like particles, I create an improper association. Expressing or not expressing that they show up as waves is incorrect. They function in their preeminent way, which could be called quantum mechanics. A particle does not behave like a weight hanging and swinging on a spring. It disdains a little more than a typical representation of the nearby planetary gathering with minor planets moving this way and that. It no longer looks like a cloud or haze, including the middle part. There is nothing that you have seen before.

There is, in each case, an unravelling. In this way, electrons continue basically like photons. They are both crazy.

The way they act requires a great deal of imaginative personality to estimate them, as we will all outline something different from anything we know. No one knows how it is."

The physical reality proposed by quantum speculation is the information we have about a system and the examinations we make based on this information. According to Bohr, these assumptions made in our psyche have nothing to do with external reality. Simply put, our inner world had nothing to do with the natural, external world, which was the essential enthusiasm of Aristotle's physicists to this day. Physicists abandoned their old thoughts about this view and agreed that quantum understanding is only our understanding of the physical picture.

The material world we can see exists only as data in our minds. As it were, we can never have direct encounters with the matter in the rest of the world.

Jeffrey M. Schwartz, neuroscientist, and professor of psychiatry at the University of California, portrayed this end from the Copenhagen translation:

John Archibald spoke: 'No wonder is a wonder until it is a wonder looked at.'

Amit Goswami expanded on this result:

Suppose we ask: Is the moon there if we do not look at it? To the extent that the moon is, finally, a quantum object (made up altogether of quantum objects), we must say no, says physicist David Mermin.

Perhaps the most significant and slippery presumption we consider in our youth is the material universe of articles out there, paying little attention to who the eyewitnesses are. There is evidence of this suspicion. For example, if we look at the moon, we will discover where we anticipate it should be along its traditionally determined direction. We venture that the moon is constantly there in space-time, anyway, when we are not looking. Quantum material science says no. If we are not looking, the impetus of the moon's possibility spreads, but by a limited amount. The moment we look, the wave stops in a fraction of a second; thus, it could not be in space-time, and it bodes well to receive an otherworldly philosophical hypothesis: "There is no object in space-time without a conscious subject taking a look at it."

Of course, this concept applies to our world of perception. The presence of the moon is evident in the external world. However, when we take a look, all we experience is our vision of the moon.

Jeffrey M. Schwartz framed these lines 'The Mind and the Brain' about the reality shown by quantum material science:

The work of perception in quantum material science cannot be overstated. In traditional material science [Newtonian physics], the observed pictures are independent of the psyche that observes and inspects them. In quantum physics, however, a physical quantity has real value only through an act of observation.

Schwartz also summarized the views of various physicists on this topic:

As Jacob Bronowski wrote in 'The Rise of Man':

"One of the goals of the natural sciences was to provide an accurate picture of the material world. One of the achievements of twentieth-century physics has been to show that this goal cannot be achieved."

Heisenberg said that the concept of objective reality was 'so fuzzy'. In 1958 he wrote that 'the laws of nature which we formulate mathematically in quantum theory no longer have to do with the particles themselves, but with our knowledge of elementary particles.

CHAPTER 7

The Law of Attraction and Quantum Physics

T he Laws of Attraction have become quite common words nowadays, especially when it comes to improving one's life. TV commercials, movies, print media, and songs have become typical. However, the motivation behind the law and its implementation are two different things. The writing is full of understandings and clarifications about the reason for the law, but too little has been done to describe the material science. Much has been learned about the meaning of the law and its application, but the world is still waiting for the mechanics to show how to benefit from it beyond simply maintaining optimistic thoughts. What are the processes that make it work? Our efforts have focused on developing and designing resources that allow individuals to apply the law more elegantly, effectively, and with less effort. We discovered a missing link on how to use and execute this incredible fundamental theory of magnetism.

An examination of quantum mechanics showed that the demonstration of looking at reality produces it. Striving to recognize something allows nothing to manifest. Likewise, if you do not know something, it does not exist in your abstract reality. The supposedly misleading impact demonstrated that positive or negative practices would create comparative effects. Research by Dossey et al. has indicated that the petition affects whether the collector knows about it.

It is becoming progressively apparent that we are co-creating our reality in the way we think and feel. In other words, in our unique and personal understanding of reality. We draw into ourselves what we hold to be true.

There is only a quantum of energy as a flow of possible positions and movements in space. This is omnipresent but, at the same time, null. Before something happens to make them "true," there is no light or matter, but what is it? Research has shown that it is something called wavefunction collapse. Further investigation has revealed that it is consciousness, which performs this action that manifests reality.

The wave function includes all potential outcomes of a given situation, but only one occurs in the physical world when a participating consciousness collapses. For now, scientists are all peering over each other's heads as they look at the screen of the latest supercollider at Cern, Switzerland. They expect two protons to collide, and when they do, the screen shows an "episode," highlighting the release of all the subatomic particles in them. What they don't realize is that their observation on the screen causes wave functions to appear. Through witnessing what we want to manifest in our minds through imagination, inner listening, and a clear sense of our thoughts, we will learn to build what we want.

Similarly, we can resist breaking down possibilities into objects, events, and circumstances we do not desire by refusing to give them our attention. A consciousness must experience a packet of energy to be effective. Until then, it lies shrouded on the other side of the quantum veil in the mystery of possibility. The "wave function" collapses when the energy is detected, making it measurable in the real world. Such particles cannot be seen with the naked eye, but only through advanced devices that can reveal where the particle was, and the velocity and position detected at that moment.

Bohr once said, 'It is wrong to think that the task of physics is to find out what nature is like. Instead, physics is about what we can say about nature."

After the most fascinating and sensitive experiments that the human mind has ever been able to develop over 80 years, there are now no opinions about quantum physics that have been conclusively and scientifically demonstrated. Nor can any objection be raised to the conclusions of the experiments conducted. Scientists have tested quantum theory in hundreds of ways and have received the Nobel Prize on several occasions for their work.

Matter, the most basic concept of Newtonian physics and once unconditionally regarded as absolute truth, has been removed. Materialists, proponents of the old belief that matter was the only building block of existence, were confused by the lack of matter suggested by quantum physics. Now all the laws of physics had to be explained in the realm of metaphysics.

The shock this caused to materialists in the early 20th century was far greater than expressed in these lines. But quantum physicists Bryce DeWitt and Neill Graham describe it:

"No development in modern science has affected human thought as profoundly as the advent of quantum theory." Torn from secular patterns of thought, physicists of one generation were forced to confront a new metaphysics. The difficulties that caused this reorientation to continue to this

QUANTUM PHYSICS

day. Thus, it was that physicists suffered a severe loss: "their grip on reality." Since both consciousness and spirit exist in the spiritual realm on the other side of the veil, it makes sense that we can control our reality even better when we are in an inspired state or a good mood. Once we are connected to Origin, we have more power in our lives and throughout our sphere of influence to create real, constructive change. Once we stay in that state of grace and reverence, trends arise, synchronicity increases, and people want to be around us and want to join us in that state of perfection. A positive wave function can collapse more easily while in that mode.

To stay in that state requires a complete concentration of consciousness. The immediate knowledge of the divine, the one and all that leads to belief in the world, leads to the confidence needed to ascend to a higher-order level. We need to understand how to drop only the positive and helpful wave capacities, removing our focus from those that cause harmful impacts. The idea, speech, and conduct of a solitary individual can change the course of the worldwide web; all things considered, this wonder of "looking" at reality regularly applies to the following tangible tools. The moment you hear something from inside your inner voice, real-world soul or higher self, or feel something from somewhere inside your own heart, do something very similar when you see it ... you experience it. Experience is, along these lines, a better word to portray the genuine, entire presence of the activity of a wave breaking. Also, incredibly better when discernment can be multi-tactile. Seeing, hearing, and encountering an encounter leads to a condition of tactile reverberation, where all tangible cycles are facilitated. A healthy autonomous sensory system adds deep relaxation and enormous innovativeness simultaneously to an established encounter (see the tactile reverberation hypothesis).

Each of us has a section that is creaturely and otherworldly. We are not one of these things, but rather an awareness in between, whose job it is to choose between the options offered by these two essential limiting components of ourselves and defeat the contention produced by the inconsistency between the polarities - remember the old children's programs where a sweetheart shows up on one side with a bit of friend on the other? That's life.

When a person creates his positive spiritual development and growth, he attracts a positive experience into himself. But when a person takes the wait and see the approach of evolution, sits back, and "goes with the flow," taking whatever is presented to them, they fall victim to the "strange attractor" or chaos of nature taking itself apart and requiring rearrangement and

QUANTUM PHYSICS

reassembly at a higher level of order. The heart does not rule the world and our lives; they are inherently disordered ... in chaos. However, the four fundamental forces that physicists claim materialize spirit have recently been redefined as celestial attractors that, over time, establish patterns of order. This is the real mystery of how reality expresses itself. Space is the original force that creates the universe from a point or singularity of zero dimensions. Time is used to concatenate the ends into the other attractors.

An individual must become an attractor to use the Law of Attraction. However, there are four orders of attractors:

1. The 1st order attractor Point - leads to being drawn into a specific task or being trapped in a rut by being too focused on an idea or fixation.

2. The 2nd Order Cycle attractor - allows you to get trapped in a logical thought or endless cycle that essentially repeats itself repeatedly.

3. The Tori 3rd Limit Request Attractor is a positive development in that it considers an intricate progression of Energy yet is somewhat obligatory due to its semi-periodicity.

4. The fourth-order attractor Odd is the chaotic behaviour of the Absolute that undoes all that is not considered in order.

The activity's accentuation, reason, coherence, and articulation pay to choose the attractor they fall. The best game plan is to adapt to the Torus attractor, direct it and compose one's entire life field in the imperatives of that dynamic topological space, which limits it and allows it to turn out to be more intermittent and repeatable, allowing the combination of ordered turbulent change constrained by an equivalent measure of determinism. An individual can transform himself into an unlimited toroidal attractor by bringing in what he needs rather than taking whatever the peculiar turbulent attractor has to bring to the table. This is the polished nature of the Law of Attraction and the key to its understanding and execution.

Look at an image in the spectral domain of an object in the time domain through open-eye meditation and observe such attractors in actions that reveal secret desires or intentions beyond the threshold of consciousness. These images are deliberate palettes in which purpose can be visualized. It has been shown that all geometries and shapes come from different ratios of real and imaginary partials that combine into the wave function's complex structure. Knowing this, and that spoken, or breathed, speech is nothing more than a flow of these natural and imaginary elements, it becomes theoretically possible to learn how to draw what you want to see on the screen, hear what you want to hear in your headphones, feel what you want to feel inside, learn how to interact with and direct the behaviour of the automata that make up all things.

The first step in learning to collapse the wave function is to sense and experience what you think and feel inside. This, in turn, is the result of applying the law of attraction by studying how to connect with the ideal attractor while paying attention to images of it—utilizing a two-stage multi-tactile approach that requires the client to turn into the specialist of their change, establishing the distinction and pressure between the administrator and the machine, an end that can also be depicted as player versus a tool, emotional versus goal, tangible versus potential, or as specialists know it, genuine versus non-existent. These natural and non-existent parts are called robots and are the least complex and most inseparable components or construction squares of presence. We are at the same time administrators and cadres at the most basic degree of nature. The change of then will allow us to set up the second and forever regulate their lives. Thus, the exemplification and ultimate destiny of breaking the wave system realize the interpenetration between actors and instruments.

CHAPTER 8

The Double Light Experiment

Imagine an "electron cannon" that shoots electrons toward a wall with two holes (or slots) that are equal distance (D) from it, and equal distance (D') from the centre of the wall (Figure 1.0). The electron gun is mounted on a turret, which moves back and forth from side to side, much like an oscillating fan. Given this movement, it is clear that we are not aiming electrons at the holes; instead, they are fired in a very random fashion. However, the caves themselves are the same size and large enough to let an electron through.

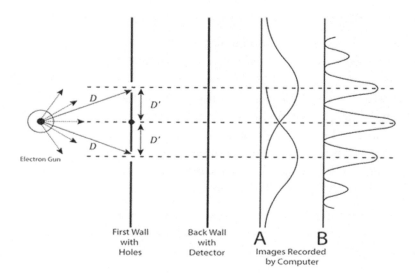

Figure 1.0. An "electron gun" fires randomly at a wall that contains two holes that are the equal distance (D) from it a similar distance (D') from the centre of the wall. The electrons are stopped at the back edge, where a detector records their positions and sends this information to a computer. Image A is the distribution we get when we put sensors next to each hole to observe an electron passing through. Here, we see no interference pattern, and in fact, we get the results we predicted for an electron acting strictly as a particle, in which case it simply passes through one hole or the other. However, image B is the distribution we get when no detectors are present. Here, we see a pattern of interference when electrons pass-through holes.

As the electrons head toward the holes, some of them will pass through, and some will not. The passing electrons continue their way until they end up hitting another wall located much lower down that acts as a backstop. At this backstop wall, the final position of each electron is recorded by a detector, which then sends this information to a computer for further processing.

As we continue to terminate more and more electrons (we need to get large sizes), more and more electrons pass through and hit the back splitter. From developing all the positions of many electrons, the PC can make an example of diffusion. If our measurements are adequate, at that point, from this scattering, we get expertise with the probability of finding an electron in a given situation on the back divider when it arbitrarily ends at the two openings. All things considered, what does scattering resemble?

Before we acquire it, let's take a moment to try to anticipate the results. If an electron behaves strictly as a particle, we will reasonably expect it to pass through one hole or the other. Also, an electron passing through a hole will either "slam" into the side or edge or pass through it unscathed. If it gives straight through, we will find it directly behind the hole - at the "centre," so to speak - when it hits the back wall; while it is bumped, we will find it hitting some distance farther on either side of the centre. We anticipate that the distribution for a given hole will be the maximum number of hits that occur directly at the centre with all of this in mind.

In contrast, the number of hits steadily decreases farther away from there. Finally, the distribution will be the same on both sides of the centre. In other words, it will be symmetrical.

Okay, we have a pretty clear picture of what we're going to see. But let's experiment and find that the resulting distribution on the computer screen is nothing like what we had imagined. Instead, we find a distribution with the maximum located between the two holes - it's not even in the centre of either hole! The distribution is still symmetrical on each side of this maximum (at least there is that), but we don't see the steady decrease in the number of hits that we imagined as we move away. Instead, on both sides, we find peaks where the number of hits is high, and then from these peaks, there is a steady decline to zero, where not a single electron appears. What has happened?

Well, in our foresight, we assumed that an electron behaves like a particle, but we should have known better since all quantum particles exhibit wave-particle duality. In short, the distribution formed by the set of positions of many electrons shows an interference pattern. Earlier, we talked briefly about how interference can occur between waves. Therefore, there must be waves associated with our electrons that are causing this interference pattern. What are they? Recall that the quantum probability determines the position of each electron on the back wall, as we mentioned earlier. In turn, the wave function gives the quantum probability (the absolute square of); it seems that we have found the "wave" that is causing the interference.

Let's try to recognize this in more detail. Instead of firing many electrons all at once toward the holes, we fire only one electron at a time. Initially, we notice that shortly after shooting an electron, it arrives at the back wall, and its position is detected. So far, so good. However, as we continue to shoot single electrons at the holes, we notice something very peculiar. In the end, we end up with the same interference pattern that we saw before when we were shooting lots of electrons. In other words, it doesn't matter if we shoot several electrons at once or one at a time; the same interference pattern appears! This means that a single electron, upon encountering the two holes, ends up interfering with itself.

This seems so strange that we decide to do one last experiment to get to the bottom of things. Next to each hole, we put a detector that will record an electron passing by it. Surely this will shed some light on the strange results we are getting. Again, we shoot one electron at a time toward the holes, over and over, until we can see the distribution on the computer screen. We find that the interference pattern has disappeared, and we are left with the distribution of electron positions that we originally predicted! In other words, when we are not looking at the holes (with our detectors), a single electron experiences the interference. Yet, when we look, we find that the electron passes through one hole or the other, and the interference pattern disappears completely.

These experiments illustrate the very essence of quantum mechanics. We see that an electron behaves like a particle when it hits the back wall and is detected by the detector as a localized entity. Yet, somewhere in the middle, there is interference due to its wave nature and its "interaction" with both holes simultaneously. This wave nature is intimately related to the quantum probability of finding the electron at a particular location on the back wall, which ultimately leads to the shot distribution we see. If we try to determine exactly where an electron will end up on the back wall to see which of the holes it goes through, the whole thing falls apart, and the interference disappears altogether.

Even though we chose to experiment with electrons, all quantum particles exhibit this kind of weird behaviour. If all this sounds more like science fiction than actual science to you, know that you are not alone. The physical consequences of quantum mechanics are plain weird compared to our everyday experience.

A GAME OF PROBABILITY

Quantum theory and quantum mechanics have shaken the very core of our understanding of the physical world in which we live. Energy, light, atoms, and matter - all of these significant factors have come under scrutiny. Even the very concepts we use to describe some of them, such as "wave" and "particle," have been pushed to their breaking point, forcing us to accept the existence of wave-particle duality for all quantum entities (electrons, light, and the like). And as if all this were not enough, determinism itself, which had always been a part of classical physics, now had to be abandoned for uncertainty and a global quantum probability. Unfortunately, these latter notions often cause the most significant confusion.

Heisenberg's uncertainty principle defines a strict physical restriction (imposed by nature) on how much we can know about specific pairs of variables, such as the position and momentum of an electron along a given direction. In other words, having better measurement devices will never reconcile this inherent uncertainty or deepen our knowledge. It also means that a quantum particle, such as an electron, does not have a well-defined trajectory. It moves along. Instead, it "moves" between quantum states according to quantum probability, which is related to the Schrödinger wave function. Bohr's atom with its "jumping electrons" - for all that it lacks - illustrates this quite well, much to Schrödinger's chagrin.

While the quantum probability is reminiscent of the classical possibilities of Maxwell and Boltzmann, nothing could be further from the truth. The latter are self-imposed to ease the burden of complicated mathematics; they still retain the underlying determinism so dear to classical physics. In contrast, the quantum probability is an unequivocal statement by nature against the determinism of the whole. Indeed, recklessly abandoning the well-oiled "world machine" for nothing more than a "game of chance" is arguably the most significant challenge to one's sensibilities. Nevertheless, by all accounts, quantum mechanics, with all its "weirdness" and probabilistic nuances, has stood the test of time.

CHAPTER 9

Emission of Black Bodies

The light wave hypothesis of light was the general hypothesis of light during the 1800s. This hypothesis was captured by Maxwell's conditions and surpassed Newton's corpuscular theory. Nevertheless, the hypothesis was challenged by clarifying hot radiation, electromagnetic radiation emitted by objects dependent on their temperature. So, how can one test or distinguish friendly radiation?

Similarly, with some other quantum tests, the ability to make a fruitful test depends on accessible identifiers and means to measure and conduct the problem itself.

For example, researchers can test for hot radiation by installing a device to distinguish the radiation from an object depends on a particular temperature, denoted by T1. People living and breathing emit radiation in every way, so to have the ability to quantify it properly, shielding must be used, so the radiation is analysed as a limited pillar. By using shielding, a scientist can create the desired conditions to focus a narrow beam. Therefore, a scientist can begin to create a suitable environment for this experiment.

To make this narrow bar, a researcher uses a dispersive medium, for example, a crystal, located between the body or object, which delivers the radiation and the radiation indicator. This promotes the scattering of radiation frequencies at a point. At that point, the indicator decides on a particular range or edge, basically the limited pillar. This pillar contemplated a representation of the total power of electromagnetic radiation through all frequencies.

Yet, how do we interpret all the power through the frequencies, and how could we decrease the different qualities to make functional conditions?

So we should clarify a couple of central issues. To comment on is that the power per unit of a frequency trait is understood as radiance. The mathematical documentation teaches us to decrease the different qualities to zero and make the following condition: $dI = R(\lambda)\, d\lambda$. Using the crystal, a researcher can recognize dI or the absolute strength over all frequencies, so you can characterize the radiance for any frequency by working backward through the condition. We should currently look at how we can fabricate some information base for frequency versus radiance curves.

Remember that each information base is processed through an assortment of investigations. Understand that researchers routinely analyse, repeatedly, constructing a heap of information that structures several extensions. When working with these extensions, you can begin to produce a superior understanding of how much radiation will happen from a particular element, but in addition to how extreme it will be at some random temperature.

For example, one may suspect that the absolute transmitted power multiplies as the temperature increases or decreases. However, when we consider the frequency with the most extreme radiance, we find that the opposite happens; that is, at that particular frequency, the force will go down as the temperature increases. Consequently, as the temperature rises, frequencies may change their radiance strength, but the overall radiance strength will continue to expand with time.

So, if the temperature goes down, at that point, the most incredible power of a single frequency will go up. The general or absolute control of the item will go down relative to the temperature. For this situation, temperature takes on a necessary function in how the test results will play out. Expanding or decreasing the temperature can affect the results. However, researchers inevitably expect specific effects. So, if they increase the temperature, however, they get unexpected results than what they were Expecting, the researcher will look for any possible flaws in the arrangement of the process itself. In any case, with every effort, they assemble an extraordinary storehouse of information about how these different parts connect.

Once again, we go consequently to how to ascertain when light reflects on infinite things. How would you do the edge, making sure you measure your little tree precisely?

A simple strategy to do this is to stop looking at the light. Instead, please focus on the item that doesn't discover it. Light removes itself from articles; however, researchers will accomplish this test by recognizing a black body or writing that doesn't show any light by any stretch of the imagination. Something else, the analysis goes into an attempt to skip what is proven.

Finishing this test binds a case, if conceivable, of metal, with a small opening. If or when light beats on the beginning, it enters the crate; however, it no longer jumps out. As an answer, the hole, not the container, is the black body of the test. Any distinct radiation outside the aperture is evidence of the

QUANTUM PHYSICS

extent of radiation in the crate. Researchers examine this evidence to sympathize with what is happening inside the container.

The critical thing to note is that the metal box is spent to stop the electric field at each mass of the crate, initiating a hub of electromagnetic vitality at each partition. Hence, the stationary electromagnetic waves are controlled inside the crate.

Secondly, the number of endless waves with their constant frequencies within a characterized go, containing a condition that brings the crate volume back to equilibrium. By investigating the standing waves and then following this condition, one may well venture into three measurements. As we have reported with the measures, numerous speculations take growth from three actions into some others, but they may not have reliable testing alternatives.

Third, traditional thermodynamics brings an essential truth: radiation in the container is in warm harmony with the dividers of the matter at a specific heat. The radiation inside the container is assimilated and retransmitted by the walls, regularly delivering the oscillation inside the radiation event. These swaying particles' warm and unique vitality is quietly consonant oscillators, so the typical kinematic dash coordinates the average possible spirit. Consequently, each wave gives the entire radiation life in the container.

Fourth, the thickness of vitality is related to brilliance. The vitality thickness is represented as the vitality per unit volume within the relation. The estimation of this is first order from the measurement of radiation passing through a segment of the surface region with a depression.

As encapsulated by the Rayleigh-Jeans equation, excellent materials science neglected to anticipate the actual effects of these analyses, basically because superior materials science ignored representing shorter frequencies. At more extended frequencies, the Rayleigh-Jeans equation concurred all the more firmly to the recognized information. This disappointment was spoken of as the luminous calamity. In the mid-1900s, this was a massive problem as it raised doubts about such unique ideas as thermodynamics and electromagnetism as an element of such equality.

Verisimilarly, this is where quantum material science is deduced into play. Almost, Max Planck devoured quanta to create what could be set up as a free lot of solidarity. Consequently, quanta would be proportional to the recurrence of vitality. With this hypothesis, no vertical wave could have more stamina than kT, and subsequently, a high radiation event would be surmounted, deciphering the luminous demolition. At last, the recurrence assigns the dynamism of each quantum were a similar standard.

Although this came from a condition that appropriates the analyses' information correctly, it was not as shocking as the Rayleigh-Jeans detail. This recipe built the initial phase of quantum material science as we probably know it today. Einstein even confined it as the focal norm of the electromagnetic region, while Planck had, in the past, utilized it only to answer the topic of solitary analysis. Although it took researchers a while to get used to what is currently known as Planck's constant, it is presently regarded as a fundamental aspect of quantum material science or quantum mechanics.

This was just one aspect of the large group of analyses that characterize quantum material science. Another early examination is the collaboration with wave-molecule duality, a test that was known as photoelectric impact.

QUANTUM PHYSICS

CHAPTER 10

Resistance

In the eyes of electricity, not all materials are created equal. Some materials are quick to allow the flow of charges, while others do their best to impede the flow of electricity. Commonly, a material's ability to conduct heat is a good indicator of its willingness to process electricity. Metals are, in general, great conductors in terms of heat and electricity. The opposite is true for water, wood, glass, and air. There is a neat phenomenon that can prove this.

When two surfaces are in contact, friction can cause a transfer of charges and thus a build-up of potential difference. For example, woollen clothing rubbing against the ground commonly causes such an imbalance (especially on cold, dry days). As long as you don't touch anything conductive, the extra charges stay on your clothes and do nothing. The air around you is not favourable enough for them to move, despite the potential difference. But touch a metal handle and - zap - the charges flow, leaving you slightly shocked.

How can we define electrical resistance mathematically?

Suppose we create a particular potential difference inside an object. Then, in a material with little electrical resistance, a relatively strong current will result. However, the same potential difference will only result in a weak current if the electrical resistance is high. So it makes sense to define electrical resistance as the ratio of the potential difference V to the current I:

$$R = V / I$$

At an absolute value of V, a large current will lead to a small amount of resistance, while a small current will do the opposite. This is exactly what we wanted to reflect. Materials with low electrical resistance are called conductors (metal) and those at the other end of the scale insulators (water, wood, and so on).

Let's go back to the fluid flow analogy. As mentioned, we can interpret voltage as the pressure difference that will cause flow. The current is then equivalent to the velocity of the flow. What does that make the electrical resistance? In this context, we can interpret it as friction. A high value of resistance does to electrical discharge what a rough surface does to fluid flow. It hinders the formation of a strong current (flow velocity), given a potential difference (pressure difference).

On a microscopic scale, electrical resistance results in part from collisions within the material. A potential difference generates an electric field throughout the conductor, which exerts a force on the charges, usually electrons, causing them to accelerate. Their movement ends quickly and abruptly when they collide with an atom in their path. Because the electric field is still active, the charged particle will accelerate again and collide with the atom in its path. This is the electric flow on the most miniature scale. It is not a continuous gentle flow of electrons but rather a series of rapid accelerations and violent collisions. Of course, the more atoms there are and the larger they are, the harder it is for electrons to flow at a high average speed (producing a high rate of charge flow = current).

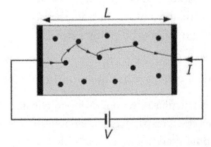

(Visualization of the path of an electron through a conductor. Not a straight line, but a series of accelerations and collisions)

A word of warning: electrical flow can cause serious injury and even death. If you are working with electricity and are not one hundred percent sure of what you are doing, stop immediately. Even a seemingly harmless 7.5 W vacation light can burn or kill a person if handled carelessly. Never try to repair electrical appliances unless you are a professional. About 60 people die each year in the United States for this very reason. If a device develops smoke, becomes unusually hot, or smells "burnt," turn it off immediately and do not turn it back on until a professional has taken care of the problem.

Stewart-Tolman Effect

American physicists Dale Stewart and Richard C. Tolman have created an interesting experiment that is able to demonstrate the nature of voltage in an easy-to-understand way. Before moving on to the investigation, imagine a truck carrying a particular gas in a container. Initially, both the car and the gas are at rest, and as is the case with all bodies in the universe, they tend to remain in their current state of motion. However, when the truck driver presses the gas pedal, a force causes the truck to accelerate.

The gas in the container is not directly affected by force. It remains at rest for the first few milliseconds. Only when the back wall of the container begins to push it forward does it accelerate as well. This causes gas to accumulate in the back. As long as the truck is accelerating, there will therefore be a pressure difference inside the container. Once the truck assumes a constant velocity, the pressure equalizes.

In an electrical conductor, charges (the usual electrons) are present even with a potential difference or current. They are just sitting there, waiting to be moved by the presence of an electric field. If you accelerate the conductor, this gas of electrons behaves no differently than the gas in the previous example. The acceleration causes the electrons to accumulate in the back. If you attach a voltmeter to the conductor at this point, it will show the presence of a potential difference. Once the conductor goes into a state of constant velocity, the electric potential quickly disappears.

This so-called Stewart-Tolman effect, the appearance of voltage when accelerating a conductor, shows that a non-uniform distribution of charges causes a potential difference. If there is nothing to maintain it, the resulting current will balance it. The experiment also "demystifies" electricity by showing that charged particles obey the same mechanical laws as all other matter. When you are uncertain about how to interpret voltage, think back to this experiment.

Piezoelectricity

In 1880, brothers Pierre and Jacques Curie made a curious discovery that would lead to many great applications in the years to come. They discovered that when certain materials (mainly crystals) are deformed, a potential difference is created. This intriguing behaviour is called the piezoelectric effect. Shortly after the discovery, Nobel laureate Gabriel Lippmann suggested that it be possible to reverse this effect. The Curies quickly demonstrated that this is indeed the case: when a voltage is applied to such materials, they deform due to it.

To explain this behaviour, we need to look at the symmetry centres of the positive and negative charges in the crystal. In an unstressed state, the centres coincide, and the crystal appears neutral. However, when the crystal is deformed, the centres shift and move apart. This leads to a net charge and, therefore, a potential difference. In the process, the mechanical energy used to deform the object is transformed into electrical energy. The effect can be used for pressure and acceleration sensors, as both cause deformation of a body. A notable example is the use of piezoelectric sensors in microphones. These can pick up pressure fluctuations caused by sound waves and transform the sound into the electrical flow. The ability to reverse the effect allows us to build devices that convert the electrical discharge into pressure waves and sound (speakers).

An unusual application of piezoelectricity can be found in a Rotterdam nightclub aptly named "Watt." The owners have built the dance floor on many crystals that compress and expand as visitors dance the night away. These deformations create enough electricity to power the club for most of the night.

Lightning

What comes to mind when you think of electricity? For most people, the answer is lightning. Lightning strikes are rare opportunities to observe electrical flow in action, which is usually hidden from us (for a good reason) using insulators and enclosures. Let's take a quick look at this spectacular natural phenomenon.

On hot days, the sun's rays significantly heat the ground, which raises the air's temperature near the bottom. Given the right circumstances, this can lead to convection: air parcels become buoyant and grow. As they do so, their temperature drops, and once it falls below the dew point, the water vapor inside the parcel condenses. A cloud is born.

The formation of a cloud is associated with a separation of charges. Observations show that the top of the cloud ends up positively charged, while negatively charged particles dominate the bottom. Yet, no theory can fully explain this electrical structure. Several ideas have been proposed that could provide an answer and are currently being researched. We will not go into the mechanisms for further discussion. However, it is sufficient to know that this separation of charges leads to a potential difference within the cloud and between the cloud and the ground.

There is a natural tendency to level out any inhomogeneous distribution of charges through the electric flow. This is not possible in the case of a cloud because the air acts as an insulator. The leaders "want" to flow but cannot because of the high electrical resistance of the surrounding air. However, once the potential difference reaches a critical value (about 50,000 volts per meter), a low-resistance channel of ionized air forms and lets the charges through.

Now the dam is broken, and the resulting current is enormous. Billions of electrons and positive ions race at speeds of 360,000 mph through the channel. This causes the temperature to rise almost instantaneously to 22,000 K and higher. The rapid thermal expansion of the lightning channel causes thunders so typical of lightning. We commonly think of lightning as going from the clouds to the ground. However, the opposite is also possible and frequently happens at skyscrapers and in the mountains.

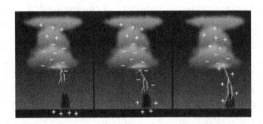

Thunder travels at about 340 m/s, the speed of sound, while light from the electrical flux reaches you almost instantaneously. This provides an easy way to estimate the distance of the thunderstorm. When you see the flash, start counting seconds; when you hear the thunder, stop. Then, divide the result by three - this is the approximate distance in miles (divide by five for the space in miles).

When lightning strikes a person, the electrical flux can cause severe injury and even death. In the United States alone, there are, on average, 35 lightning fatalities per year, most of them males (74%) between the ages of 20 and 59. If a thunderstorm is approaching, go to a haven (building, car). Never seek shelter in sheds, boats, or under trees. Do not bathe or shower and use the phone only in an emergency. If there is no shelter, go to a low spot away from trees, fences, and poles. Tingling sensations or electrified hair are signs of immediate danger. In this case: crouch down and make yourself as tiny a target as possible.

CHAPTER 11

Photoelectric Effect:

Einstein's Theory

hen electromagnetic radiation of appropriate frequency is made to strike the surface of a metal, such as, say, sodium, electrons are emitted from the metal. This phenomenon of electrons being emitted from certain materials (which include various metals and semiconductors) by electromagnetic radiation is called the photoelectric effect. This effect can be demonstrated and studied with the help of a set-up like the one shown in Figure 1.1.

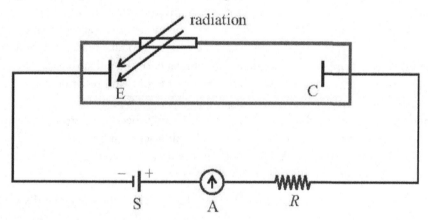

Figure 1.1: Set-up to study and analyse the photoelectric effect; E is the emitting surface while C is the collecting electrode; A is a current measuring device; S is a DC voltage source whose polarity can be reversed; R indicates a resistor; the actual circuit may not be as simple as shown here.

An emitter metal electrode (E) and a collector electrode (C) are enclosed in an evacuated chamber in which a window admits electromagnetic radiation of appropriate frequency to fall on E. A circuit consisting of an EMF source (S), a resistor (R), and a sensitive current meter (A) is established between E and C. The polarity of S can be changed so that C can be either at a higher or lower potential relative to E.

Photoelectric emission characteristics

This arrangement can be used to record different excitatory characteristics of the photoelectric emission. For example, suppose that C's potential (V) toward E is positive for a given intensity of incident radiation. Then, all electrons emitted from E are collected by C, and A records a current (I). This current remains nearly constant when V is increased because all photoelectrons are collected by C whenever V is fluxed. This is known as the saturation current for a given intensity of incident radiation.

However, this whole phenomenon of registering a current due to the emission of photoelectrons from E depends on the frequency (v) of the radiation. If the frequency is sufficiently low, photoelectric emission does not occur and no photocurrent is recorded. For the moment, we accept that the frequency is high enough for photoelectric emission to take place, and refer to Figure 1.1. Suppose the frequency and intensity of the radiation are held constant. In that case, we now reverse the polarity of S and record the photocurrent with an increasing magnitude of V. We find that the photocurrent persists but gradually decreases until it becomes zero for a value V =-The magnitude (Vs) of V for which the photocurrent becomes zero is called the stopping potential for the given frequency of the incident radiation. This is shown graphically in Fig. 1.1.

The lower of the two curves shown in Fig. 1.2 describes this change in I with V for a given intensity (J1) of the incident radiation. The frequency v also held constant at a sufficiently high value.

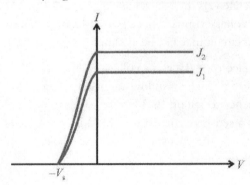

QUANTUM PHYSICS

Figure 1.1: Graphical representation of the characteristics of photoelectric emission; the variation of photocurrent I with applied voltage V is shown for two values of radiation intensity, J1, and J2 (> J1), while frequency ν is held constant; the stopping potential Vs. is independent of power.

If, now, the experiment is repeated for some other value, say, J2, of the radiation intensity, then a similar variation is obtained, as in the upper curve of Fig. 1.1, but with a different value of the saturation current, the latter being higher for J2 > J1. However, the stopping potential does not depend on the intensity since, as seen in the figure, both curves give the exact value of the stopping potential.

On the other hand, if the test is repeated with different frequency values, keeping the intensity fixed, we find that the stopping potential increases with frequency (Figure 1.2). It turns out that if the frequency is decreased, the stopping potential reduces to zero at some finite value (say ν0) of the frequency. This value of the frequency (ν0) turns out to be a characteristic of the emitting material and is referred to as its threshold frequency. No photoelectric emission from the material under consideration can occur if the frequency of the incident radiation is not higher than the threshold frequency. Furthermore, for ν > ν0, photoelectric emission takes place for arbitrarily small values of the intensity. The effect of lowering the intensity is to decrease the photocurrent without stopping the emission altogether.

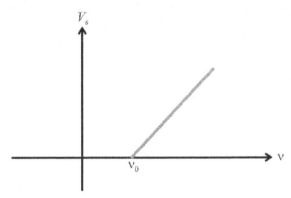

Figure 1.2: Variation of the stopping potential with frequency; no photoelectric emission occurs if the frequency is less than the threshold value ν0, no matter how large the intensity may be.

The role of photons in photoelectric emission

All these observed features of photoelectric emission could not be explained by classical theory. For example, classical theory tells us that whatever the frequency, photoelectric emission should occur if the intensity of radiation is high enough because, for high intensity of radiation, electrons inside the emitting material should receive enough energy to escape, overcoming their binding force.

Einstein first gave a complete account of the observed characteristics of the photoelectric effect by invoking the idea of the photon as a quantum of energy, as introduced by Planck in connection with his derivation of the black body spectrum formula.

While photons in blackbody radiation were the energy quanta associated with standing wave modes, similar considerations apply to propagated radiation. Indeed, the electric and magnetic field components' strengths of propagating monochromatic electromagnetic radiation vary sinusoidally with time. Again, a mode of field propagation can be viewed as a quantum mechanical harmonic oscillator of frequency, say, ν. The minimum value for which the energy of the radiation can increase or decrease is again $h\nu$, and this increase or decrease can likewise be described as the appearance or disappearance of a quantum of energy, or a photon, of frequency ν. Such a photon associated with a progressive wave mode also carries a momentum like any particle, such as an electron (in contrast, an energy quantum of blackbody radiation has no net velocity). The terminologies for the energy and momentum of a photon of frequency ν are the de Broglie relations we are now familiar with:

Where λ stands for the wavelength of the propagating monochromatic radiation and where only the momentum magnitude has been considered.

When a monochromatic radiation of frequency ν is made to engrave on the surface of a metal or semiconductor, photons of the same frequency interact with the material, and some of the exchange energy with electrons in it. This can be interpreted as collisions between photons and electrons, where the power of the photon engaged in a crash is transferred to the electron. This energy transfer may be sufficient to drive the electron out of the material, which is how photoelectric emission occurs.

Bonded systems and bonding energy

Metal or semiconductors is a crystalline material in which many atoms are arranged in a regular periodic structure. The electrons in such material are bound to the entire crystal structure. In this context, it is essential to grasp the concept of a bonded system. For example, a small piece of paper glued to a board constitutes a bound system, and it takes some energy to rip the piece of paper off the board. If the power of the network consisting of the paper separated from the board is taken as zero (in the process of energy accounting, any energy can be given a pre-assigned value since power is indeterminate in the measure of an additive constant), and if the energy required to tear the paper is E, then the principle of conservation of energy tells us that the power of the bound system with the paper glued to the board must have been -E since the tear energy E added to this initial energy gives the final power 0.

As another example of a bound system, consider a hydrogen atom of an electron "glued" to a proton by the attractive Coulomb force between the two. Again, it takes energy to get the electron out of the atom, resulting in an unbound electron separated from the proton. The power of the split system, with both the proton and electron at rest, is taken by convention to be zero, in which case the expression gives the energy of the hydrogen atom bound with the electron in yet another steady state. Note that this energy is a negative quantity, which means that positive energy of equal magnitude is required to tear the electron away from the proton. This method of pulling an electron from an atom is known as ionization. It can be accomplished with the help of a photon, which provides the necessary energy to the electron, and the process is called photo-ionization.

In the same way, a hydrogen molecule is a bonded system consisting of two protons and two electrons. Looking at any one of these electrons, one can deduce that it is not bound to any one of the two protons but the pair of protons together. The two electrons are shared by the team of protons and form what is known as a covalent bond between the protons. Again, it takes some energy to get any of these electrons out of the hydrogen molecule.

The minimum energy needed to separate the components of a bonded system is called the binding energy. When they receive this amount of energy, the components separate from each other, without gaining any kinetic energy in the separated configuration. If the bound system gets more incredible energy than the binding energy, the extra amount goes to generate kinetic energy in the components. In this context, an interesting result concerns the situation where one of the components is much lighter than the other. In this case, the extra energy is consumed almost entirely as kinetic energy of the more lightweight component.

When I speak of a bound system, I tacitly imply that it must be seen as a system made of two components. However, the same system can also be seen as a system made up of more than two components. For example, in the piece of paper glued to the board, the components I have in mind are the paper and the board. But given a sufficient supply of energy, the board can also be broken into two or more pieces, and then one would have to think of a system consisting of more than two components. The board and the piece of paper consist of many molecules, and the molecules can all be torn apart from each other. Similarly, all of the two electrons and two protons that make up the hydrogen molecule can be pulled away from each other, so a different amount of energy would be needed than would be required to get a single electron separated from an ion. The latter we call the electron binding energy in the hydrogen molecule.

CHAPTER 12

Quantum Reality

n isolated quantum system is always in a superposition state, a state of potentiality before being observed. It has no intrinsic properties except charge and mass. For example, position and velocity do not apply to a quantum system before measurement because size produces properties.

Isolated quantum objects, such as electrons, behave like ideas: they exist in an abstract mathematical space. What is observed of these abstract objects during the measurement process is their expression/projection in our measurement apparatus; it is the apparatus that plays the primary role in quantum physics.

The expression we get on the apparatus is the representation of the quantum object in terms of the capacity of the measuring device. If the instrument measures the position of quantum objects, then using this apparatus on an electron, we get the representation of our electron in position space. Thus, the position we assign to an electron is only its expression in our apparatus and not an objective property of the electron itself - in the same way that the letters of the word apple do not belong to the idea "apple."

Just as it makes no sense to assign letters to an idea, giving a position or velocity to an electron itself is equally meaningless. This is the starting point of quantum physics. According to quantum theory, what has been thought of as objective reality has no actual, independent existence; it is a mere expression in the observer that plays the role of a representative basis.

Philosophers have divided the qualities of an object into primary qualities and secondary qualities.

The primary qualities are defined as those that are independent of any observer; whether the object is lost in deep space or observed, these primary qualities are unaffected because they exist in the object itself. Examples include shape, size, and, according to classical physics, even the object's state of motion, including position and velocity.

The secondary qualities are defined as those that depend on the observer; these qualities exist whenever the object is observed or experienced in some way. In other words, Secondary qualities exist in our experience of the object and not in the thing itself. Examples include colour, taste, smell, apparent size and shape, etc. Colour, for instance, does not exist in the object; the object is made of atoms and electrons that are colourless and tasteless; what we experience as colour comes from our brain's interpretation of the frequencies of light emitted by the object's atoms.

Although the secondary qualities do not exist in the object itself, the object must be present for us to experience its secondary qualities. Thus, these qualities are neither in the thing nor in the subject but belong to and exist in the interaction interface. They appear only during the event in which the subject interacts with the object.

What has happened with the advent of quantum physics is that our world's primary, objective qualities have turned out to be only secondary qualities. Based on experimental findings, quantum physics has shown that qualities such as position, velocity, energy, and thus the state of motion of an object do not exist in the object itself; they belong to the interface of interaction, which is now the measuring apparatus, i.e., the representative basis. As a result, quantum physics has undermined the idea of realism.

This paradigm shift that undermines the objectivity of our world is what makes quantum physics counterintuitive. Once we accept that an isolated quantum object possesses no specific position or velocity or energy, then we cannot even imagine such a remote system - not so much because we don't know what it looks like, but rather because a quantum object is not a type of object that looks anything at all, much less resembles anything. The quantum object has no appearance, and therefore is not something to look at!

The properties of quantum objects, and those of objective reality, essentially quantum mechanics, exist only during observation and experience. The world exists only in our experience. The belief in the independent existence of the world is itself something within the background.

The world does not possess any objective properties because the world is not a type of thing that includes something. Instead, the world is only an appearance.

Quantum particles - like electrons, protons, and other subatomic particles - are only mathematical constructions, like Plato's regular solids. Thus, a measuring device shows only a few figures, which man must interpret as something meaningful.

QUANTUM PHYSICS

The world of phenomena, the world we know, is not something we have experienced; it is not something apart from experience; it exists only within observation and experience. We perceive the world as an entity but what appears on the interface of interaction. Werner Heisenberg writes in his Physics and Philosophy:

"More than being a simple description of the subatomic world, quantum physics is a statement about the classical picture of the world. It exposes the inconsistencies and paradoxes inherent in our everyday view of the world as something material, deterministic, and indifferent to consciousness."

For example, Heisenberg's uncertainty principle rejects the possibility of material solidity.

Now let's see how the superstitious ideas of material solidity and constitution came about and why false.

The classical worldview is rooted in our conscious experience. Thus, for example, we experience objects as solid, liquid, or gaseous; therefore, we have these main distinctions in classical physics.

The objects of our experience seem to enjoy an independent existence subject to certain causal relations; hence, the objective, deterministic aspects of the classical view. But if we pay close attention to the same experiences within which these ideas first arose, we realize the following:

All objects, and our whole world, are first given to us in experience, but the experience is a temporal flow of consciousness; the solid aspect of things is itself something experienced in time. The solidity is itself a phenomenon that belongs to the material flow of experience. (If I do not use the phrase my experience, it is because the "I" is something experienced; it belongs to the content of experience and not outside it. Experience is not something I have because I am something experienced.)

The world and its objects are phenomena constituted in the temporal flow of subjective experience. It is in time that they are what they are. Objects are first and foremost material entities because in and through time, their existence has meaning as flux. In other words, objects are, first of all, extended in time before they are developed in space. It is in the primordial river of time that reality makes its first appearance and acquires its objectivity.

It is essential to distinguish between something that flows and something that exists within the flow. For example, the flow of a river is the flow of water molecules; these molecules exist even when they are not flowing like the river; the flow is only one possible state of the movement for them.

However, the temporal flow of experience within which objective reality is constituted is essentially different from flow in the sense of a river. The material flow is a constant state of flux within which everything is comprised; nothing exists outside of this flow because existence is itself constituted in time.

The fact that reality is essentially a temporal phenomenon is not calculated in our classical view of reality. The uncertainty principle in quantum theory is a technical restatement of the non-physical but secular constitution of phenomena. This is why a point particle does not make sense in this primordial flow that we call Heraclitan flow.

An analogy can help us see how something can exist only within a flow: imagine that you see from a distance something that seems to be a circle; when you get closer, you see a bright circular object suspended in space. Although a circular object exists, the glowing ring is nothing more than the trace of a fireball being spun very fast by an invisible fire dancer.

The circular object that appears to exist as something in space is a mere appearance that exists only within the flow of fire created by the dancer. If the rotation stops, the circle no longer appears. Notice that the loop never existed; in the beginning, it was just a fireball that occurred as a circle because of a constant flow.

Now, if you hit such a circle with a stick, it will lose its shape and disappear altogether because you have disturbed the very flow within which it existed as an appearance. This is very similar to what happens in the quantum world when the measurement is performed. The act of measurement destroys the very appearance under consideration.

We see that some of the strange features of quantum physics are not so strange if we correct our classical worldview by reflecting deeply on the experiences within which we have the world in the first place. For example, suppose objective knowledge must begin with experience, end with experience, and never lose its grounding in experience. In that case, our minimal ignorance about the vital role of experience in the constitution of reality leads to knowledge detached from reality.

The facts of the microscopic world articulated in quantum physics are experiential references to the underlying reality as Heraclitan flux, which is essentially a temporal-subjective reality rather than a spatial-objective speculative fiction.

CHAPTER 13

Holography and Quantum Gravity

The myth of gravity

My older son reads a primary science reading that explains gravity. He says, "Dad, look, every time I jump in the air, a force pulls me down to Earth, do you know why? Ironically, I said no. He went on to explain to me that it is simply because of gravity.

Gravity pulls everything toward the Earth, and all masses attract, he added. Finally, he went on to ask me why gravity behaves the way it does. I replied, son, this is just the way nature works. Newton, in response to a similar question, said, "non-fringe hypothesis." This translates as "I have no idea."

Surprisingly, such a straightforward question about a common phenomenon that we experience every minute of our lives has not been answered directly for many centuries, even by the brightest physicists. So why does gravity exist in the first place? What is gravity? Commonly, it is defined as an attractive force that all matter possesses. However, matter attracts another issue, and the strength of the attraction depends on two things, the mass of the objects and the distance between them.

Gravity and its subordinate attractive energy originate from the Latin word gravitas, from gravis, meaning heavy. It has been explained that the advanced meaning of fascination did not manifest itself until the time of Newton. For researchers like Galileo and Copernicus, until the beginning of the 20th century, the word gravity had no concrete meaning. We get from old-fashioned material science that attraction or gravity is a characteristic wonder in which items or bodies with mass attract each other. Attractive energy is responsible for keeping our planet in its circle around the Sun. It is also responsible for observing the Moon in its rotation around the Earth.

Without gravitation, we would witness chaos in the universe, as it causes scattered matter to coalesce. More important than anything, gravity is responsible for the formation of tides, natural convection, etc. For example, we learn that objects maintain their orbits in space because of the force of gravity acting on them. Yet, despite the successes of MS particle physics in unifying the three fundamental forces of nature, the fourth - gravity - remains elusive. This is perhaps due to our lack of understanding of its origin and why it is there in the first place. Of all the natural forces that exist, gravity is the force we are most familiar with, but its origin has eluded even the most brilliant physicists.

Newton's physics explains how an apple falls toward the Earth. Einstein took it a step further by explaining gravity as the deformation of the fabric of space-time. So far, all of these theories have described how gravity works, but their flaw remains in their inability to explain how it arises. For example, Newton had many reservations and doubts about the action of gravity at a distance. For centuries, this question, among others, has not been satisfactorily answered. For example, how can one massive object attract another at a great distance without any mediation?

Another question that has also remained unanswered is how to explain the attraction between two objects with no time of action? Recently, many publications have pointed out that gravity may not be a fundamental force at all but rather an emergent property of the deeper underlying structure of the universe. Since the 2010 publication of Verlinde's article entitled "On the Origin of Gravity and Newton's Laws," many subsequent articles have examined this idea in a new light. The overwhelming feeling from these publications is that a new worldview that emphasizes the primary role of information over matter and energy is destined to take centre stage.

However, I would first like to examine the history and implications of the theory of gravity, from the time of the ancient Egyptians to Einstein and recent attempts by several researchers who believe that the universe may be emergent and holographic. I will also examine how this new model, in which gravity might not be a fundamental force, might change our picture of space-time, boosting the physics of computation and information - what we now call digital physics.

CHAPTER 14

Unified Energy and Unified Matter

This brings us to the concept of Unified Field, Unified Force, Unified Matter, and Theory of Everything.

In this part, we will better see that the "energy string" is the fundamental entity of the universe. These energy strings are the unique hybrid of mass and energy from which all types of particles and forms of energy are created.

I also believe that the five types of energy strings provide most, if not all, of the properties found in the universe.

Therefore, this leads me to believe that the concepts described above will eventually lead us towards the "Unified Field Theory."

Unified Field

In molecular material science, a unified field hypothesis is a drive to characterize each primary power and connections concerning elementary particles into specifications of a hypothetical singular structure. Thus, fields that arbitrate the relationships between disjoint things can be named in material science. James Clerk Maxwell transmitted the underlying field hypothesis into his way of thinking about electromagnetism during the nineteenth century. In the early twentieth century, Albert Einstein set up general relativity, the field hypothesis of attractive energy. As a result, Einstein and others attempted to create a field hypothesis bound together in which electromagnetism and gravity would emerge as various parts of a solitary main field. Unfortunately, they failed, and to this day, gravity remains a past attempt at a bound together field hypothesis.

At subatomic separations, fields are represented by quantum field hypotheses, which apply the thoughts of quantum mechanics to the whole earth. During the 1940s, quantum electrodynamics (QED), the quantum field hypothesis of electromagnetism, was entirely resolved. In QED, charged particles associate while producing and holding photons (tiny packets of electromagnetic radiation). As a result, we are exchanging photons in a round of subatomic "getting." This hypothesis works so well that it has become the model for speculations of different powers.

Furthermore, I said that the "field" is just the flow of vitality strings. Consequently, if all "fields" are types of vitality strings, at that point, we can make a "Carried Field Theory" considering "carrying together" these vitality strings.

Expressly, I accept that these vitality strings are varieties of an essential type of vitality string. I see the kinds of vitality strings recorded and represented above as branches of a tree. We know we have comparable components. We should locate the central part of the tree.

The moment we distinguish that central part, we will, at that point, could bring all the fields together into a complete picture.

Unified Matter

These strings of energy also bring us to Unified Matter. Many scientists, including Heisenberg, talk about not only a unified field but a "Unified Matter." Heisenberg believed that the quarks he discovered had similar properties, thus leading us to an understanding of a Unified Matter.

We can do this with our energy strings. The energy strings I have proposed are the hybrid of energy and matter. Therefore, by unifying these energy strings, we will not only unify our energy fields, but we will also unify our point.

The future of the unified field based on energy strings

The future of the Unified Field, the Unified Force, and the Unified Matter will all be developed from the five basic types of energy strings that I listed and mentioned above.

For the first time, we have a common language and standard structure for each type of field, force, and all the properties observed in each particle. For the first time, we have commonalities where each field, energy, and particle can resemble the other areas, forces, and particles. This is a big step.

Now we can take these commonalities and find the True Unifying Concept behind them all. In fact, from one Universal Energy, we can create all things. All energies, matter, movements, and observations can be explained based on these few concepts.

These secrets are being discovered now. They will be revealed gradually through a future series of publications, eventually leading to the Unified Energy Solution and the Theory of Everything.

CHAPTER 15

SUPERCONDUCTORS

NO RESISTANCE

hen does resistance disappear? The answer to this question came from Kamerlingh Onnes already in 1914. He proposed a very ingenious method to measure resistance. The experimental scheme seems simple enough—coil lead wires in an omitted cryostat - an apparatus for performing experiments at low temperatures. The cooled helium coil is superconductive. In this case, the present flowing through the coil, creating a magnetic field around it, which can be easily detected by the deviation of the magnetic needle located outside the cryostat. Then close the key, so that now the superconducting shot has been short-circuited. However, a compass needle has been deflected, indicating the presence of current in the coil is already disconnected from a current source. Observing the arrow for several hours (until the helium evaporated from the vessel), Onnes had not noticed the slightest change in the direction of the deviation.

According to the experiment results, Onnes concluded that the resistance of the superconducting lead wire was at least 10/11 times lower than its resistance in the normal state. Subsequently conducting similar experiments, it was found that the current decay time exceeds many years, indicating that the superconductor resistivity is less than 10 25 ohm - m. Comparing this with the resistivity of copper at room temperature, 1.55 x 10 -8 ohm-m - the difference is so significant that one can safely assume that: the resistance of the superconductor is zero, it is challenging to call another monitor and change the physical quantity that would be faced in the same "ground zero" as the resistance of the conductor under the critical temperature.

Familiar recall from the Joule - Lenz school physics course: current I flowing in the conductor with a resistance R generates heat. At this consumed power P = I 2 R. As little resistance to metals, but often limits the technical possibilities of various devices. Heated wires, cables, machines, appliances; thus, millions of kilowatts of electricity are thrown to the wind—heating limits the transmission power, the power of electric cars. Therefore, in particular, is the case with electromagnets. Getting solid magnetic fields requires a large current, which leads to the release of a considerable amount of heat in the windings of the solenoid. But the superconducting circuit remains calm, and the current will circulate without damping - zero impedance, no power loss.

Since the electrical resistance is zero, the exciting current in a superconducting loop will exist indefinitely. The electric current, in this case, resembles the current produced by the electron's orbit in the Bohr atom: it is like a vast Bohr orbit. The persistent current and the magnetic field generated by it cannot have an arbitrary value; they are quantized in such a way that the magnetic flux penetrating the ring assumes values that are multiples of the elementary quantum flux $F_{on} = h / (2e) = 2{,}07\ 10\ 15$ Wb (h - Planck's constant).

Unlike electrons in atoms and other microparticles, the behaviour of which is described by quantum theory, superconductivity - macroscopic quantum phenomenon. The length of the superconducting wire through which persistent current flows can reach many meters or even kilometres. Thus, the carriers describe a single wave function. This is not the only macroscopic quantum phenomenon. Another example is superfluid liquid helium or neutron stars substance.

Electric resistance superconductors

No experimental method is fundamentally valid to prove that any value, particularly the electrical resistance, is zero. It can only be shown to be less than a specific value determined by the accuracy of the measurement.

The most accurate method of measuring low impedance is to measure the decay time of the induced current in the closed circuit of the test material. A decrease in the energy of the current time $LI^2/2$ (L - the rate of the inductance circuit) consumed by Joule heat:

hereby integrating (I_0 - value of current at $t = 0$, R - circuit resistance).

The current decays exponentially with time, and the electrical resistance determines the attenuation rate (at a given L).

For small R, the formula can be written as:

Here dI - change in current during Δt.

Experiments using a thin-walled superconducting cylinder with minimal values of L showed that the superconducting current is constant (accurately) within a few years. It followed that the resistivity in the superconducting state is less than 4 x 10 25-ohm m or more than 10 17 times lower than the resistance of copper at room temperature. Since the possible decay time is comparable to humankind's existence, we assume that R DC in the superconducting state is zero. Therefore, it is easy for us to deduce at this point that the superconducting current is nothing but a real example of perpetual motion on a macroscopic scale!

When R = 0, the potential difference V = IR on any segment of the superconductor, and hence the electric field E inside the superconductor is zero, the electrons that create a current in the superconductor move at a constant speed without being scattered by the thermal vibrations of the atoms in the crystal lattice and its irregularities, note that if E is not equal to zero, the electrons carrying the superconducting current are accelerated without limit. Thus, the current could reach an infinitely large value, which is physically impossible.

The situation changes if the superconductor applied to the variable potential difference creates a variable superconducting current. During each period, the current changes direction. Consequently, the superconductor must exist in an electric field, which periodically slows down the superconducting electrons and accelerates them in the opposite direction. Because it consumes energy from an external source, the electrical resistance of alternating current in a superconducting state is zero. However, since the mass of the electrons is very small, the power loss at frequencies below 10 10 - 10 11 Hz is negligible.

TUNNEL EFFECTS

In 1962, a paper appeared by an unknown author, B. Josephson, which theoretically predicted the existence of two extraordinary effects: stationary and unstable. Josephson theoretically studied the tunnelling of Cooper pairs from one superconductor to another through any barrier. Before proceeding to the first Josephson effect, we briefly summarize the tunnelling of electrons between two metal parts separated by a thin dielectric layer.

The tunnel effect has been known in physics for a long time as it represents a typical problem of quantum mechanics. The particle (e.g., an electron in the metal) approaches the barrier (e.g., a dielectric layer) to overcome what classical ideas cannot because its kinetic energy is insufficient. However, the barrier zone with its kinetic energy could very well exist. On the contrary, according to quantum mechanics, passing the barrier is possible. The particle may have a chance, so to speak, to pass through the tunnel into a classically forbidden region where its potential energy would be more like a full, that is, classical kinetic energy since it is negative. Quantum mechanics for microparticles (electron) holds the uncertainty relation $\Delta h \Delta r > h$ (x - coordinate of the particle, p - its momentum). When a slight uncertainty of its coordinates in a dielectric $\Delta h = d$ (d - thickness of the dielectric layer) leads to considerable uncertainty, its impulse $Dp \geq h / \Delta x$, and consequently, the kinetic energy $p\,2\,/\,(2m)$ (m - a mass of particles), the law of conservation of energy is not violated. Experience shows that between two metal electrodes separated by a thin insulating layer (tunnel barrier), the electric current can flow more.

JOSEPHSON EFFECT

The physical objects in which the Josephson effect takes place are now called Josephson junctions or Josephson elements. To imagine the role played by Josephson elements in superconducting electronics, one can draw a parallel between them and the p-n junctions of semiconductors (diodes, transistors) - the fundamental element of conventional semiconductor electronics.

Josephson junctions are weak electrical communications between two superconductors. This connection can be made in several ways. The most commonly used types in counterproductive link practice are:

1) tunnel junctions, in which the bonding between the two superconducting films is made through a thin insulating layer (tens of angstroms)- SIS- structure.

2) "Sandwich" - two superconducting films interacting through a thin layer of a standard metal between them (hundreds of angstroms)- SNS structure.

3) The bridge-type structure is a narrow superconducting bridge (bridge) of limited length between two massive superconducting electrodes.

Superconducting "carriers" at $T = 0$ K are all conduction electrons n (0) (electron density). When the temperature increases appear elementary excitation (normal electrons) so that the concentration of n s of the superconducting electrons at a temperature T:

n s (T) = f (0) -n n (T),

where n n (T) - the electron concentration at a normal temperature T. In the Bardeen-Cooper-Schrieffer (BCS) for T → T c (critical temperature)

p s (T) ≈ Δ 2 (T)

Where 2 D (T) - the width of the energy gap in the superconductor spectrum. All superconducting electrons form associated state pairs, known as electron Cooper pairs.

The Cooper pair combines two electrons with opposite spins and pulses and therefore has a net-zero spin. Unlike normal electrons, which spin 1/2 and thus obey the Fermi-Dirac statistic, Cooper pairs follow the Bose-Einstein statistic and condense to one, the lower energy level. A characteristic of Cooper pairs is their relatively large size (about 1 micron) is much larger than the average distance between teams (of the order of interatomic spaces). Such strong spatial overlapping pairs mean that all (condensation) of the Cooper pairs is coherent, described in quantum mechanics as the wave function of a single: W =. DELTA.E ix. Here A - amplitude of the wave function, the square of which characterizes the concentration of Cooper pairs, h - the phase of the wave function, i - imaginary unit, P - -1. In the case of normal electrons, which are fermions, the Pauli exclusion principle, the energy of electrons is never exactly equal. Therefore, the Schrodinger equation for these particles follows that the phase velocity dq/dt of the wave functions of the normal electrons differ; thus, phase h is uniformly distributed in the trigonometric circle sum over all particles explicit dependence on h disappears.

There is a weak electrical connection between the superconducting electrodes due to the poor superposition of the wave functions of the Cooper pairs of the electrodes, so this contact is also superconducting. Still, the critical current value density is much (by several orders of magnitude) lower than the critical current density of the electrodes j c ≈ 10 8 A/cm 2. For tunnelling and sandwich-type structures at Josephson critical current density, junctions-ing is typically in diapazonej j jc from 10 1 to 10 4 A/cm 2, and their SB area within modern technology can be made from a few hundred to a few microns square. Therefore, the critical current of the Josephson element I c = j jc. S can be from a few milliamperes to a few microamperes.

CHAPTER 16

Evolution of Elementary Particles

I believe that the universe was initially created not as elements of a larger nucleus but as elementary and sub-elementary particles. An electron and a proton attracted each other, forming the primitive hydrogen atom. At one stage, hydrogen atoms fused through a fusion reaction, creating heavier elements.

Contrary to the big bang theory, what happened at the time of the big bang 13.7 billion years ago, or perhaps earlier; rather than the entire universe coming into being, only two particles of opposite and equal momentum exited the light speed reference frame along the axis of motion of the sonic reference frame. Their exit was in opposite directions so that one particle was in the order of movement of the reference frame while the other was in the opposite direction. The equal and opposite moments ensure that no disturbance (recoil) occurs at the exit point. Because sonic is inertial from the point of view of this reference frame, the measurements of mass, time, and length are finite as calculated from within this sonic frame. All laws of physics are official within this sonic system, including the laws of conservation of energy and momentum. The exit of the two particles from the sonic system is implemented according to the laws of conventional mechanics as considered from within this framework.

The laws of energy and momentum are conserved at the time of exit.

An observer is hypothetically located at a resting place to monitor and measure the sequence of events of the two particles as they exit the sonic reference frame to their final destinations. The hypothetical resting position is assumed because the very first exit of the particles occurs in zero space. Thus, before the particles exit, the observer at rest will not detect the two particles when they are in the light speed reference frame. According to the theory of special relativity, an observer at rest sees the sonic reference frame and its contents as a point in space with zero volume and infinite mass. When the two particles leave the light speed reference frame, the observation spot detects and finds that the forward and backward particles have faster and slower speeds than the speed of light. Both particles left the light speed reference frame in opposite directions: one forward (faster) and the other in the reverse order (slower). Both particles are now visible at the resting place of observation because they have accurate velocities propagating in the same space, as explained above.

After the two particles leave the sonic reference boundary, they apply a push to pull each other along, trying to connect to return to their ground state, which is the speed of light. Since the two particles are currently in the subsonic and supersonic reference contours, they cannot re-emerge from the light speed boundary. Their re-emergence is taboo because their masses increase as they approach the speed of light again. Heavy groups need a truly extraordinary vitality to get to the sonic speed they need. Unable to resurface the sonic reference profile, the two particles arrive at a certain distance from each other and then stop. At this position, they have exhausted most of the viability they got from the sonic reference profile. The supersonic molecule is currently at supersonic speed, while the subsonic molecule has arrived at relative rest. The moving supersonic molecule remains at a reasonable fixed distance from the subsonic molecule while maintaining its supersonic velocity. The foremost opportunity for the supersonic molecule to maintain a fixed separation and its speed is around the subsonic molecule. In the case where we indicate that the supersonic molecule is the electron and the subsonic molecule is the proton, at that point, we can announce the introduction of the hydrogen iota. The hydrogen iota is framed with the proton at the focal point of a circular path modeled by the electron moving with supersonic velocity.

The previous arrangement of the occasions accomplished by the two particles after their takeoff from the sonic reference edge can induce three significant ends: (1) In their drive to recombine back to their actual state, it is nothing but the electrostatic fascination between the proton and the electron. (2) The supersonic mass revolving around the subsonic mass is nothing more than the arrangement of the hydrogen particle with the electron revolving around the proton. (3) Their fascination stops at a fixed separation limited by their expansion in mass, which characterizes the nuclear sweep. In the resting place of observation, an observer detects the formation of a hydrogen atom in an inflated space. From the point of view of the observation post, the newly formed hydrogen atom and the increased space come from nowhere because the sonic reference frame is an abstract place observed at rest. This may answer the controversial question, 'Can something come from nothing?' The answer is yes; something can come from an unobserved place.

The two particles that have come out are now the electron and the proton. Today, the question is, "Why did we decide that the electron is the supersonic mass and the proton is the subsonic mass? The answer is simple: since the electron, because of its supersonic velocity, is constantly moving and has never been found at rest, it must be the supersonic particle. The proton can be found at rest; therefore, it must be the subsonic particle.

The evolution of space and the simplest element, the hydrogen atom, is the first step towards the formation and development of the universe. The author believes that hydrogen was the first constituent of the cosmos. Consequently, the evolution of the hydrogen atom and space is the first step toward the universe's expansion.

The following is a formal presentation of the theory of evolution of elementary particles.

POSTULATES

To continue with the turn of events and the development of elementary particles, one could embrace the assumptions that accompany the theory:

1. Inertial reference contours exist at any speed, including slower, faster, or the speed of light.

2. Within any inertial reference frame, life is typical, and all laws of material science are substantial.

3. Masses can leave the reference contour of the speed of light at faster or slower speeds. However, they cannot return to that speed (exit in only one direction).

As indicated by old-style material science, inertial reference envelopes can exist at any speed. Accordingly, the one proposed follows traditional Galilean and Newtonian material science. The hypothesis of special relativity did not rule out the probability that an inertial reference edge could exist at the speed of light. This hypothesis only expressed that such a reference edge could not accelerate at that speed and that such a reference edge could not be identified from rest. This hypothesis of special relativity reveals that a sonic reference envelope could exist at the speed of light with the requirement that this reference contour existed initially at that speed and did not accelerate from slower or faster than the speed of light. Then again, no law of material science rejects the presence of a reference contour beyond the speed of light. Along these lines, postulate one could be accepted.

Postulate two is in line with all classical laws of physics as long as the frame of reference is non-accelerating or inertial.

Postulate three is consistent with the Lorentz mass transformation equation, which can be rewritten as:

$$m = m'\sqrt{1 - \frac{v^2}{c^2}} \qquad (C.6')$$

Where:

m' is the mass in the sonic reference frame at the time of departure,

m is the mass measured in the subsonic or supersonic reference frame, is the particle's velocity, while c represents the speed of light.

The equation states that as the velocity of a particle moves away from the speed of light (faster or slower), the particle's mass, m, is reduced to a smaller gathering, measured at rest. Thus, the reader can easily verify that a particle cannot fall within the speed of the light reference frame because its mass becomes infinitely large, according to the equation.

As a result, particles can easily escape the speed of light in either direction, up or down. Furthermore, because of its nearly infinite mass, when the velocity of a particle increases toward the speed of light, that particle is prohibited from re-entering that speed. As a result, groups can easily transition from sonic to subsonic or supersonic speed, while the reverse is not permitted. In general, leaving the sonic reference frame is a one-way process. Therefore, postulate three is valid and can be adopted.

EVOLUTION OF THE HYDROGEN ATOM.

The previous discussion of the ejection of the two masses of equal momentum bears a striking resemblance to the creation of the electron/proton pair. The supersonic and subsonic ejected particles are the electron and proton, respectively, because of the following:

QUANTUM PHYSICS

1. As observed from stationary, the ejected supersonic particle is an imaginary negative term. The negative time can be related to the electrostatic charge of this mass. The supersonic particle resembles the electron because of its continuous motion.

2. The mass of the ejected subsonic particle is positive and genuine and attracts the supersonic abundance; therefore, it must be the proton.

3. The imaginary term associated with the supersonic mass could be related to the difference between the consistency and magnitude of the two ejected groups.

4. The attraction between the negative supersonic mass and the positive subsonic mass, in their effort to return to the ground state, is evidence of the attractive electrostatic forces of the two ejected groups.

5. The two evolved masses cannot merge because such an action requires that the created masses fall within the sonic reference frame, which is forbidden according to the third postulate.

It can be generalized that the two ejected masses are no more than one electron and one proton. Their attraction up to a certain distance and the continuous movement of the electron around the proton is strong evidence for the formation of the hydrogen atom. The fixed radius between the electron and proton explains why the electron does not fuse with the proton in the hydrogen atom, regardless of the strong electrostatic attraction forces.

Let us now modify our previous spaceship hypothesis. Suppose that a sonic mass splits into two groups due to a new action in the sonic reference frame, in the other dimension, or perhaps in a black hole. The two masses gain energy to split and exit their sonic home with equal and opposite momentum along the direction of motion of the sonic reference frame. The two split particles are the electron and proton that form the first hydrogen atom.

CHAPTER 17

Mathematics, the Language of Physics

The Role of Mathematics in Physics

Let's pay a little attention to the following two comments:

-1. Roland Omnès, an esteemed professor of theoretical physics in France and a graduate of the elite Ecole Normale Supérieur in Paris, states at the beginning of his 2002 "Quantum Philosophy":

"If I had to name the greatest thinker of all time, I would say, without hesitation, Pythagoras, who lived on the Greek island of Samos, 6th century B.C.) .... He said that numbers rule the world".

-2. To the aspiring scientist or mathematician:

"Follow the numbers, no matter where they take you. They contain the truth of the Universe."

The crucial working relationship between the world of physics and mathematics is both synergistic (they work together or enhance each other) and symbiotic (two different systems live together). Moreover, this relationship is a crucial operating characteristic in the progress of science.

First, note that there are essentially two branches of mathematics: Theoretical Mathematics and Applied Mathematics. This discussion will focus on Applied Mathematics. But, in deference and respect to the theoretical mathematician, I would like first to explain what a theoretical mathematician does.

Examples of where applied mathematics relates to the world.

Many people in the world are trying to figure out what is going on in the physical world. For example, there are:

- Weather and climate analysts try to understand what makes and drives different times and climates.

- Oceanographers who try to understand currents and tides in the water.

- Engineers who construct buildings and calculate what weather forces can or will bring down a structure.

- Aeronautical engineers who determine the strength of materials.

- NASA engineers are designing the trajectory of space flights to the planet Mars.

- An automotive engineer is designing a new type of contour (curvature) for the hood of a new sports car.

- Physicists are analysing a substance when the substance's temperature approaches zero degrees Kelvin or analysing how electricity flows within the importance, even near zero degrees Kelvin.

- Computer architects are trying to reduce the number of components (molecular or smaller) on a memory chip or logic board.

- Physicists are trying to understand the inner workings of black holes.

- Physicists are developing string theory.

- Physicists are trying to incorporate Quantum Gravity into the Standard Model.

- Physicists are trying to harness and apply quantum physics to a wide range of objects, including computers, communication systems, medical sensing, monitoring systems.

- And thousands of others.

These are all people who rely on the mathematics to describe the process they are analysing. That's why this area of mathematics is called Applied Mathematics.

QUANTUM PHYSICS

First, they try to understand the process. Then, when they can, they try to describe it with mathematical formulas, called equations. Sometimes the procedures are known and available; sometimes, there is no formula general. In the latter case, they try to derive it independently or wait for another group of mathematicians to develop the formula. The development of the mathematical equation often provides the developer with additional insights into the physical process they are studying.

Mathematics is usually an independent profession performed by mathematicians; they produce mathematical formulas. However, because not many people "buy" mathematical formulas, mathematicians work in, or with, other professions such as being instructors and professors in colleges and universities, and in government and non-profit organizations (where mathematical studies are performed), and in industrial and research organizations.

One modern application is the development of a spectacular array of mathematical formulas to describe and predict events and activities in the stock market, the bond market, the futures market, the real estate market, the credit market, the mutual fund market, and so on.

Many of us have learned about Pythagoras, a Greek mathematician (circa 500 B.C.), who observed a right triangle and applied an equation that we call the Pythagorean theorem. The theorem is that the sum of the squares of the sides is equal to the court of the hypotenuse. This equation allows the sides of a right triangle to be calculated for all right triangles of different sizes.

In the case of Pythagoras, or one of his colleagues, he saw a physical object and realized that a mathematical formula could describe the thing. (Just as Kepler did, but Kepler did it for the planets).

There are many different, well-known equations, and we often refer to them by the founder's name. For example, there is the Bessel Function, named after Frederick Bessel (1784-1846). Fortuitously, Bessel's Function can be used to describe the FM radio wave we listen to on the FM band of our car, iPod, and home radio.

And then there's an example that has always intrigued me. It has to do with one of my favourite mathematicians, Leonhard Euler (1707-1783), considered the greatest mathematician of the 18th century. 19th and 20th-century engineers built tall buildings. They would put a steel beam vertically in the ground. It was held in the environment by concrete or some other type of "construction glue." They would also attach steel beams to different beams, extending their height. Or they would place a steel beam horizontally, supported by other beams.

Leonhard Euler had developed such a formula in about 1755. This was long before steel was even thought of. So, Euler had no idea what a steel beam was. But his formula was used for a hundred and fifty years to predict the shaft's bending if or when a thrusting force acted on the beam. So civil engineers used Euler's equation to create new buildings and new cities.

The example of Bernhard Riemann's application of mathematics

Bernhard Riemann developed another set of 'exciting' equations in 1854. These equations described a different geometry from the two-dimensional Euclid geometry that you and I had learned in school. Riemann's geometry represented 'non-planar and undulating surfaces.

In the early 1900s, Albert Einstein had developed the theory of space and gravity. But he didn't know how to express it mathematically. Finally, after learning (from his mathematical friend, Marcel Grossman), he could describe his thoughts mathematically.

This application of Riemann geometry became known as a crucial part of Einstein's Theory of General Relativity. Without this mathematical description, Einstein, or anyone else, could not have made calculations about gravity, space, and the universe for various situations.

I will conclude this introduction with a quote from Albert Einstein:

"The approach to a deeper understanding of the fundamental principles of physics is related to the most intricate mathematical methods."

Three Outstanding Mathematicians: Leonhard Euler, Srinivasa Ramanujan, and John Von Neumann

I have long been aware of the extraordinary and exciting lives that many mathematicians have led - and I would like to devote many more pages to their personal histories and contributions to pure mathematics. However, since this is unrealistic for a reading dedicated to physics, I limit my discussion to only three very different mathematicians: Leonhard Euler, Srinivasa Ramanujan, and John von Neumann.

The first is famous because of his contributions in many areas of mathematics and physics, the second for his creativity and general brilliance, and the third for his creativity and leadership in three very different modern arenas.

Leonhard Euler, the Most Prolific Mathematician of All and "Master Fiddler of Strings."

The first time I learned about Leonhard Euler, the Swiss mathematician (1707-1783), was during a civil engineering course examining the strength of steel beams. The professor described an equation that he attributed to Euler.

Below I list some of his various contributions and give a number to each of Euler's "fields of endeavour." I believe, however, that many others are not mentioned here.

1. Euler was responsible for "combinatorial analysis." An example of combinatorial analysis: how many ways can you take eight different elements from a group of 21 other features? The combinatorial analysis is fundamental to developing statistical theories used in Quality Control, a vital ingredient of the modern industrial age.

2. Euler is considered the father of "modern graph theory." Euler's graphs were what are called linear graphs, which he developed to solve problems. Today we are all familiar with the "salesman's problem."

These line graphs solve the problem of where a salesperson needs to visit a certain number of customers, and the routes are analysed to minimize travel, thus, by implication, increasing sales and profits.

3. One branch of the mathematical discipline known as Calculus is "Differential Equations." There appear to have been two major phases in the development of Differential Equations. First, they are probably the most widely used system of equations to express physical situations. Second, the parameters, time and time-varying, appear in the flows of liquids and gases. In a similar time-varying case, Schrödinger used the Differential Equations to express his time-varying quantum equations.

4. In physics known as Fluid Mechanics, Euler explicitly stated the concept of internal pressure in a liquid. He also developed the equations for the formulation of a three-dimensional description used for fluid flow. The study of liquid pressure and flow is essential today in many of our standard industrial systems. Examples include the oil, natural gas, and water pipeline networks that crisscross our continent; similar pipeline networks interconnect European and Asian countries, allowing them access to products across national borders; and there are identical pipeline networks within many of the world's cities.

A CREATIVE MATHEMATICIAN AND GENIUS, SRINIVASA RAMANUJAN (1887 - 1920)

Srinivasa Ramanujan was a mathematician who had a non-standard personal history. He developed many theorems on his own, living in India, and many of these theorems were unknown in Europe and America. His work can help us understand some of the fundamental problems and unsolved theories that still challenge physicists today and will continue to do so in the future. For example, in the part of physics called 'String Theory,' it is believed that Ramanujan's formula called Modular Elliptic Function will help describe the theory further, mathematically.

John Von Neumann (1903 - 1957)

John Von Neumann (1903-1957), a Hungarian mathematician, distinguished himself from his peers, even in childhood, by having superior mental faculties. For example, he had a photographic memory. He was able to memorize and recite a page of a telephone book in a matter of minutes. Also, by the age of six, he could divide eight-digit numbers in his head. As a result, he is distinguished from being involved in many aspects of modern physics and engineering.

Von Neumann published his first mathematical paper at the age of eighteen, collaborating with his tutor, and decided to study mathematics at university. In 1926 he received his doctorate in mathematics with minors in physics and chemistry from Pázmány Péter University in Budapest, Hungary.

CHAPTER 18

The Universe

There are many aspects of computation in the universe; there are many phenomena that scientists have been able to turn into mathematical formulas. For example, to show that there is a mathematical relationship between temperature and volume, the higher the temperature of the ball, the larger its size, and because of the density, we show that it will float on the surface of the water, Archimedes used water and objects to discover this. He was able to convert this into a mathematical equation between mass and volume. Albert Einstein also managed to connect mass, energy, and velocity with a mathematical formula.

Many examples point to this. Several theories have emerged that discuss the possibility of computerization of the universe, i.e., the universe is a vast quantum computer, so what is computerization? This has always been a crucial question in the field of computer science.

In the early 1930s, computer science meant the function of the people who ran the supercomputers. And in the late 1940s, computer science was defined as: a set of steps implemented by automated computers to produce knowledge output, and this standard definition has remained for fifty years after, but now faces many challenges; While people from many fields have started to accept the idea that computational reasoning is a way to understand science and engineering.

The Internet is full of services that do computation without stopping. Researchers in physics and biology claim to discover computational processes from nature unrelated to computers; the calculation is now in all sciences. Computer scientists design, build and program computers. But again, back to our question: What can be considered a computer? What is missing from the rock (as a physical system) and owned by a computer?

According to Seth Lloyd, author of Programming the Universe, all interactions that occur between molecules in the universe transmit not only energy but also information; in other words, molecules not only collide but also perform mathematical operations, the entire universe computes each atom, an electron and an elementary particle that stores a lot of information, and every time two molecules collide, these two bits are processed.

By delving into the computational power of the universe, we can build quantum computers that store and process information at the level of atoms and electrons. This computational power forms the basis of complex systems and provides a deeper understanding of the origin and future of life.

Computer science is taught in two ways: theoretically and materially; mathematics is concerned with studying the theoretical side of computer science, providing mathematical definitions of computational topics such as algorithms, and providing theories about their properties.

One of the most important questions was: is first-order logic decidable? That is, is there an algorithm that can decide whether a particular (first order) logic sentence is a theory? Turing and Church showed that the answer is no; there is no algorithm with this description.

To prove this, they provided an accurate description of the unknown concept of a computational function and the writing of an algorithm. Turing did this through the so-called Turing machine, a virtual machine that processes separate symbols written on a tape compatible with a limited number of commands. Thus, the study of computational functions is made possible by the work of Turing and others. According to the Turing-Church hypothesis, any intuitively computable process is computable with the Turing machine; this can be formulated as follows: "Any function that is seen as naturally computable is a function that is computable with Turing."

Intuitively computable, which means that it is computable by following an efficient algorithm or procedure. The practical approach includes an expired list of clear commands for producing new symbolic structures based on ancient symbolic forms.

Some researchers study that the physical universe is a computational base, the universe itself is a computational system, and everything within it is also a computational system, where it is viewed from two different perspectives, the first of which is an automatic operator that represents the traditional computational model and the quantitative, non-quantitative model.

The idea that the universe could be a giant digital computer existed decades ago. In 1960, Edward Fredkin, then a professor at the MIT Institute Konrad Zuse, who built the first electronic digital computer in Germany in the early 1940s, proposed the idea that the universe is an integrated digital computer (and recently, the idea has been supported by computer scientist Stephen Wolfram.

QUANTUM PHYSICS

According to these physicists, the universe is a giant cellular robotic operator. The robot is a network of cells, with each section taking a state within a finite set of conditions and updating its shape in separate steps according to adjacent states.

For the universe to be cellular robotics, all physical quantities must be separated. Also, time and space must be separated. Thus, although cellular robotics can describe many basic physical phenomena, the quantum features of the universe are challenging to simulate using a conventional model such as cellular robotics.

The universe is quantum, and standard computers cannot simulate quantum systems; why? Because quantum mechanics is more exotic and counterintuitive than regular computers, in that for humans to affect a tiny part of the universe, transferring a few hundred atoms into a Second, an average computer needs more memory space than the number of bits in the entire universe, and a longer time than the current age of the universe to finish the simulation.

This is what led to the development of quantum computing models. Instead of relying on numbers - often numbers or bits - quantum computing relies on qubits. The difference between qubits and bits is that while bits can take on one value between two values: 0 or 1, qubits can take on a set of values representing the superposition of the 0 and 1 states.

According to this principle, the universe is not a classical computer but rather a quantum computer; it is a computer that does not process numbers but qubits. Furthermore, the quantum version of the universe is less radical than the conventional version since the standard version removes continuity from the universe, arguing that removing it allows classical computers to provide a literal description of the universe rather than an estimate.

The universe is a physical system subject to computation; therefore, it can be simulated effectively using a quantum computer - the size of the universe itself - and because the universe supports quantum computation. It can be fabricated using a quantum computer. It possesses computing power that is no less than that of a quantum computer the size of the universe.

We have seen how the laws of physics can be used to perform quantum computation effectively; let's find out how a quantum computer can simulate physical laws.

QUANTUM PHYSICS

Quantum simulation is the process by which a quantum computer simulates a quantum system. Due to the strange quantum properties, classical computers affect quantum systems less effectively. However, since the quantum computer is a quantum system capable of highlighting all quantum properties, it can effectively simulate other quantum systems. Each part of the quantum system that you want to emulate is stored within a group that is stored within a group of I qubits that, in turn, are within the quantum computer. The various interactions between these parts are transformed into logical quantum computing operations. The resulting simulations are so accurate that it is difficult to differentiate them from the simulated system.

No description of the universe was found as a computer before the 20th century. The ancient atomic Greeks counted the universe as a form of interaction of small parts but did not explain whether these parts were information processing units.

Laplace invented a virtual object that calculates the entire universe's future, but he considered it a separate entity from the universe and not the universe itself.

At the same time, Charles Babbage was not eager to use his device as a model of physical phenomena, unlike Alan Turing. However, the latter was interested in the origin of models and researched the subject.

The first explicit description of the universe as a vast computer was in 1956 in the science fiction novel "The Last Question" by Isaac Asimov. In this story, humans invent analogue computers to help them explore first their galaxy and then other galaxies. The connection between computer science and physics began in the early 1960s by Rolf Landauer at IBM. The idea that the universe could be thought of as a computer was proposed by Fredkin and independently by Konrad Zuse. Frieden and Zeus suggested that the universe might be a classical computer called a cellular robot containing rows of bits that interact with adjacent bits.

Stephen Wolfram has recently developed and simplified this proposal. The idea of using cellular robotics as a basis for the theory of the universe sounds interesting. However, its problem is that classical computers cannot reproduce quantum phenomena, such as quantum entanglement. Another reason, as mentioned above, is that simulating a small portion of the universe on a classical computer requires a volume equal to the size of the universe.

Therefore, it is impossible to consider the universe as a classical computer as a cellular robot. However, in his research paper: Ultimate Physical Limits to Computation, Seth Lloyd showed how the computational power of any physical system could be calculated by knowing the amount of energy in the design and the size of the system, for example: Use these limits to calculate the maximum computing power of a kilogram of material stored in a liter of volume one from space, the standard laptop computer weighs about a kilogram and takes the size of a litre hole so we will call the one-kilogram, one-litre computer a super-laptop.

A super laptop is a one-kilogram, one-litre computer (the size of a standard laptop), where every elementary particle has been put inside it for computation.

A super laptop can perform 10 million logical calculations per second at ten trillion bits. So, what could be the power of a super laptop computer?

The first significant impediment to excellent computational performance is energy. The amount of energy determines the speed range, for example: Let's take a one-bit electron, which moves here and there, whenever it has a lot of energy whenever it moves quickly here and there and can change its bit-flip state quickly, the speed at which qubits change its condition is subject to a theory known as Margolus-Levitin. The idea says that a specific physical system (e.g., an electron) varies based on its energy.

CHAPTER 19

Quantum Computing

Is Information Physical?

Computers are devices that process information. Computer scientist and physicist Rolf Landauer argued that knowledge is a part of the physical world. He elaborated this as follows: data is not a disembodied abstract entity; it is always linked to physical representation. An engraving represents it on a stone tablet, a [magnetic] spin, a [electrical] charge, a hole in a punch card, a mark on paper, or some other equivalent. This connects information management with all the possibilities and limitations of our actual physical world, its laws of physics, and its storage of available parts. If "information is physical," as Landauer put it, then it would seem necessary to treat it mechanically. In other words, the physical means by which information is stored and interpreted by computers should be considered using quantum theory. It helps to understand computation in general before addressing quantum computers.

What is a Computer?

A computer is a machine that receives and stores incoming information, processes the data according to a programmable sequence of steps, and produces the resulting information output. The term 'computer' was first used in the 1600s to refer to people who perform calculations or computations and now refers to computers that compute. Computing machines can be roughly divided into four types:

1. Computing devices for classical computational physics. These machines use moving parts, including levers and gears, to perform computations. Usually, they are not programmable but always perform the same operation, such as adding numbers. An example is the Burroughs Adding Machine of 1905.

2. Fully programmable classical mechanical electromechanical computing devices. These machines operate using electronically controlled moving parts.

They process information stored as digital bits represented by the positions of a large number of electromechanical switches.

The first such machine was built in 1941 by Konrad Zuse in wartime Germany. In theory, their programmability allows them to solve any problem found and overcome it using algebra. Thus, these were the first "universal" computers in this context.

3. All electronic, hybrid quantum-classical-physical computers. These universal, fully programmable computing machines have no moving mechanical parts and operate using electronic circuits. The first to be built was the ENIAC, designed by John Mauchly and J. Presper Eckert, University of Pennsylvania, 1946. The physical principles that describe the movement of electrons in these circuits are rooted in quantum physics. However, classical physics adequately describes how electrons represent information since there are no superposition or entangled states involving electrons in the various circuit components (capacitors, transistors, etc.). Therefore, we call these machines - essentially any computer in operation today - "classical computers."

4. Quantum computers. If ever successfully built, these devices would operate according to the principles of quantum physics. Knowledge will be expressed by the quantum states of individual electrons or other elementary quantum artifacts. In addition, there will be entangled states involving electrons in the various components of the circuit. It is expected that these computers will solve these kinds of problems much faster than any modern classical computer can.

How do computers work?

Computers store and manipulate information using a binary alphabetic language consisting of only two symbols: 0 and 1. Each symbol 1 or 0 is called a bit, short for binary digit, because it can take on two possible values. A page of text, such as the one you are reading, is represented as a long string of numbers in a computer file. A binary code represents each letter. For example, 'A' becomes 01000001, 'B' becomes 01000001, and so on.

In a typical computer, each bit is represented by the number of electrons stored in a small device called a capacitor. We can think of a capacitor as a box that holds a certain number of electrons, kind of like a container of loose grain in a grocery store that carries a certain amount of rice. Each capacitor is called a memory cell. For example, such a capacitor might have a maximum capacity of 1,000 electrons. If the capacitor is full or nearly full of electrons, it represents a bit of value 1. If the capacitor is empty or almost empty, we say it means a bit of value 0. The capacitor cannot be half full, and the circuit is designed to ensure that this does not happen. By grouping eight capacitors, each of which is either full or empty, any eight-bit number - for example, 01110011 - can be interpreted.

The role of the machine circuit is to empty or fill the various capacitors according to a set of rules called a program. Eventually, the action of filling and emptying capacitors succeeds in performing the desired calculation - for example, adding up two eight-bit numbers. In a computer, the steps are performed by tiny components of computer circuits called logic gates. A logic gate is made of silicon and other elements arranged to block or pass electrical charge, depending on its electrical environment. Logic gate inputs are bit values, represented by a full capacitor (a 1) or an empty capacitor (a 0). (The word 'gate' is associated with something going in and something coming out).

How small can a single logic gate be?

In the first all-electronic computers, such as the ENIAC, built in the 1940s, a single logic gate was a vacuum tube similar to the amplifier tubes still used today in vintage-style electric guitar amplifiers. Each box was at least the size of an inch. By 1970, the microcircuit revolution was able to reduce the size of each gate to about a hundredth of a millimetre. When things get much smaller than that, it's best to measure the length by a nanometre unit, equal to one-millionth of a millimetre. The gate size in 1970 was 10,000 nanometres.

On the other hand, a single atom of silicon, the primary atomic element in computer circuits, is about 0.2 nanometres thick. By 2012, the gates available in typical computers had been shrunk enough to be spaced only 22 nanometres apart, or only about a hundred atoms. As a result, the effective working area of the gate was less than 2.2 nanometres by ten atoms thick. This small size allows a few billion memory locations and inputs to be placed in an area the size of a thumbnail.

We are having gate sizes much smaller than this size leads to both a curse and a blessing. We leave the domain of multi-atomic physics and enter the realm of monoatomic physics. There are now variations between the principles of classical physics that sufficiently explain the average behaviour of many atoms and the principles of quantum physics that are required when dealing with single atoms. We reach a domain of random actions that does not sound good if we try to get a well-regulated system to carry out our numerical orders. A group of scientists led by Michelle Simmons, director of the Centre for Quantum Computation and Communication at the University of New South Wales, Australia, has built a gate consisting of a single phosphor atom embedded in a silicon crystal tube. This is the minor gate ever designed. This gate only works properly when cooled to a shallow temperature: - 459 degrees Fahrenheit (- 273 degrees Celsius). If the material is not as cold, the random (thermal) movement of the silicon atoms in the crystal decreases the confinement of the electron's psi-wave, which may leave the channel in which it is to be confined. For everyday desktop computers, which, after all, must operate at room temperature, this loss prevents such single-atom gates from being the basis of technology that everyone can use. On the other hand, these experiments show that, at least in theory, computers can be built on an atomic scale, where quantum physics reigns.

Can we create computers that fundamentally use quantum behaviour?

Given that physics defines the ultimate behaviour and efficiency of information transfer, storage, and processing, it is reasonable to ask how quantum physics plays a role in information technology. Since electronic computers are based on the behaviour of electrons and communication systems are based on the behaviour of photons - both elementary particles - it is not surprising that quantum physics ultimately determines the performance of information technology. But there is a subtlety here. Computer technologies currently in use do not involve quantum superposition states to represent information. Instead, they use conditions considered classical forms of material things - namely groups of electrons.

The big question is, can we create computers that use quantum states to improve our ability to solve real-world problems? If these computers were ever built, they would bypass some forms of data encryption methods much faster than any computer in operation today. This would revolutionize the field of privacy and confidentiality for computers and the Internet. For example, the encryption key, which could take thousands of years to crack, with a conventional computer could take only minutes on a quantum computer.

WHAT IS A QUBIT?

The word bit refers to both the abstract, disembodied mathematical concept of information and the physical entity that embodies the information. It is evident in classical physics that a "physical bit" carries an "abstract bit" of knowledge. There is a direct one-to-one relationship between the state of the physical bit and the value of the abstract bit, 0 or 1. We can also use individual quantum artifacts, such as an electron or photon, to embody a portion. In this case, the elementary physical entity is called a qubit, short for "quantum bit." A qubit has two different quantum states, such as the H and V polarization of the photon or the upper and lower paths. When measured, the results represent a bit value of 0 or 1. But remember that we can select different polarization measurement schemes - say, H/V or D/A. The results can then be random, with the probability of observing possible outcomes depending on which measurement scheme we have selected. In this case, there is no one-to-one relationship between the state of the physical qubit and the value of some abstract conceptual bit. Concepts from quantum physics suggest significant variations between the behaviour of classical bits and qubits. Classical bits can be copied as many times as we like, without any degradation of information; qubits cannot be copied or cloned even once, although they can be teleported. The state of the classical bit, 0 or 1, can be determined by a single measurement; any sequence of measures cannot select the quantum state of the individual qubit.

What physical principles differentiate classical computers from quantum computers?

There are significant differences between the gates used in classical computers and the gates used in quantum computers. For example, classical gates perform operations that are not reversible; understanding the output doesn't tell you what the inputs are. On the other hand, if a quantum gate works properly with qubits, it must be reversible. That is, you must be able to determine the input states by understanding the output states. This requirement arises because any quantum gate operation must be a unitary process.

QUANTUM PHYSICS

CHAPTER 20

What are your quantum thoughts?

The understanding of quantum thoughts is a set of hypotheses that suggest that classical mechanics cannot explain consciousness. However, it can explain consciousness and postulates that superposition and entanglement can play a significant role in brain functioning. Claims that understanding can filter out a movement that assigns traits to quantum phenomena such as the absence of locality and public effect.

History

Eugene Wigner developed the idea that quantum mechanics has something related to the workings of the mind. He suggested that the wave function collapses because of its interaction with all consciousness. Freeman Dyson argued that "thoughts, as exemplified by the ability to make decisions, are to some extent inherent in every electron." Philosophers and other physicists believed that these arguments were not convincing. Victor Stinger recognized quantum understanding as a "fantasy" with "no scientific basis" that "should take its place along with unicorns and dragons." David Chalmers argues against quantum understanding. He argues that quantum mechanics can be linked to consciousness. Chalmers doubts that any physics can solve the problem of consciousness.

Tactics of the quantum mind David Bohm

David Bohm saw relativity and theory as contradictory, which indicated a level in the world. Nevertheless, he retained quantum theory and emphasized that this level was suggested to represent order and an undivided totality as we encounter it, where the order of this world originates. The implicit charge proposed by Bohm applies to both consciousness and matter. He suggested that it can clarify the connection between them. He saw matter and mind in the order implied to our order as projections. Bohm argued that when we look at the matter, we find it. He mentioned the experience of listening to songs; he considered that the feeling of the changes and movements that constitute our experience of audio comes from holding the current from the mind and the past.

Penrose and Hameroff

Theoretical anaesthesiologist Stuart Hameroff and British mathematician, philosopher of science, and Nobel laureate in physics Roger Penrose collaborated to create the concept called Orchestrated Objective Reduction (Orch-OR). The two developed their thoughts and collaborated to generate.

Orch-OR since the early 1990s. They updated their vision and rechecked it. Penrose's argument is derived from theorems. In his very first reading on understanding, The Emperor's New Mind (1989), he argued that although a formal system cannot prove its consistency, Gödel's improvable results are provable by individual mathematicians. Penrose took this example to imply that mathematicians are not operating traditional proof systems but operating an algorithm. According to Xiao and Bringsjord, this line of reasoning is based on a misunderstanding.

At the very same publication, Penrose wrote: " Speculating is possible, but somewhere deep in mind, single sensory sensitivity cells should be discovered. If this turned out to be the case, then quantum mechanics would be involved in brain activity." Penrose decided that wavefunction collapse was the only potential basis for a non-calculable procedure. He suggested a kind of wavefunction collapse, which involved its reduction and occurred in isolation. He indicated that when these split, they become unstable, fall and each quantum superposition has its curvature particle. Penrose suggested that reduction does not mean randomness or algorithmic processing but rather an effect in geometry from the derived understanding and expansion. Hameroff provided a theory that microtubules are hosts to quantum behaviour. Microtubules are composed of dimer protein subunits. Dimers could include multiple electrons, each with pockets. Tubulins have other areas that include pi indole rings. Hameroff suggested that these electrons are enough to become entangled eventually and could create a Bose.

PENROSE CONTINUES

A good deal of what the mind does can be done on a pc. I am not saying that the entire activity of the reason is different from anything that can be done on a pc. I am arguing that conscious activity is something. This is not to say that understanding is beyond mathematics, although it is beyond the physics, we understand today... My assertion is meant to indicate that there must be a personality significant enough, that we still don't know about. It is not specific to our brain; it is out there in the universe. However, it performs a function and may need to maintain the bridge between the classical and quantum levels of behaviour, where the quantum dimension comes in. W. Daniel Hillis expressed his opinion, "Penrose has made the classic mistake of putting people at the centre of their world." He argues that to be so complex, the human mind must necessarily get some elixir derived from principles of mathematics, so it must involve it.

DAVID PEARCE

British Transhumanist David Pearce, guards what he calls the view of genuineness ("unrealistic crudeness attests to the fact that truth is experiential, and the conditions of science accurately portray the natural universe just like their responses"), furthermore contains the hypothesis that unitary cognitive personalities are physical conditions of quantum intelligibility (neuronal superposition). Again, as indicated by Pearce, this hypothesis is mismanaged able, contrary to numerous speculations of understanding. In addition, Pearce summarized an exploratory equation that portrays how the theory could be tested with an emission wave interferometer to find examples of neoclassical obstruction of adrenal superposition toward the onset of hot decoherence. Pearce admits that his considerations are "exceptionally theoretical, "unreasonable," and "extraordinary."

CRITIQUE

Since Penrose and Pearce acknowledge in their discussions that all of these assumptions of their quantum mind remain speculations, the ideas are not evidence-based until they make a prediction that is examined by experimentation. According to Krauss, "it's a fact that quantum mechanics is bizarre, and on microscopic scales for rapid timescales, all kinds of bizarre things happen. And we could have quantum phenomena happen that are bizarre. However, quantum mechanics doesn't change the world because you still have to do something even if you want to change things. You can't change the world by considering it." Thus, the practice of analyzing hypotheses with experiments is fraught with conceptual/theoretical, functional, and ethical issues.

Conceptual issues

The notion that a quantum impact is essential to the functioning of consciousness remains in the world of doctrine. Penrose suggests that it is crucial, but other ideas of consciousness do not mean that it is necessary. For example, Daniel Dennett has proposed multiple draft versions, which does not imply that quantum effects are essential within his 1991 Consciousness Explained. The philosophical debate on each side is not scientific proof. However, a philosophical investigation can point to crucial differences in the types of models and reveal that some of them can be observed. But since there is no consensus among philosophers, it is necessary to support a concept of quantum mind. You can find computers primarily designed to compute using the results of quantum mechanics. Quantum computation is computation using quantum phenomena, such as superposition and entanglement. They are not the same as transistor-based binary digital electronic computers. Among the most significant challenges is removing or controlling quantum decoherence. This typically means isolating the system from its surroundings since interactions with the outside world cause it to decoherence. Quantum entanglement is a body phenomenon frequently invoked for versions of the quantum brain. This effect occurs when groups or pairs of particles interact, so the quantum state of each particle cannot be described independently of another(s), even if a considerable space separates the particles. Instead, a quantum state must be elucidated for the entire system. Measurements of physical attributes such as position, momentum, spin, and polarization, conducted on entangled particles, are seen as connected.

Practical Issues

Demonstrating human brain impact through experimentation is essential. Is there a way? Could a complex electronic personal computer be shown to be incapable of consciousness? A quantum computer will demonstrate that quantum effects are intentional. Whatever the case may be, one could build quantum or electronic computers. These may explain which type of computer is capable of conscious thought. However, they do not exist, and no evaluation has been demonstrated. Quantum mechanics is one model that can offer some accurate predictions. Richard Feynman anticipated quantum electrodynamics, contingent on the formalism of quantum mechanics, "the gem of material science," for its very exact expectations of quantities like the strange attractive snap of the electron and Lamb's move of the vitality degrees of hydrogen: therefore, the model that can give an accurate estimate on understanding, which could confirm a quantum impact, is incorporated. The test is to look for an experiment that demonstrates whether the brain depends on quantum effects. It must explain a difference between a computation leading to one, which implies quantum effects, and a mind. The theoretical argument against the brain theory is that quantum states would remove up to a scale where they might be helpful for processing. Tag Mark has elaborated on this assumption. His calculations suggest that quantum systems from the mind decohere in time scales. No reactions from a reason revealed reactions or benefits. The responses are on the order of milliseconds rather than time scales.

The Penrose States

The problem with using quantum mechanics in the activity of the mind is that if it were a matter of quantum neural signs, these neural signs would disrupt the rest of the substance from reason to the extent that quantum coherence would be lost very quickly. In such a disordered environment, one could attempt to build a quantum computer. Nerve signs would have to be dealt with. But if you go down to the degree of microtubules, there is an excellent chance of getting action. For my picture, I want this action at the microtubules; the story must be an element of a scale that goes through large regions of the brain, from one microtubule to another but from one nerve cell to another. We want some coupled quantum character action that Hameroff says is happening across the microtubules. There are avenues of attack. One is that of quantum theory about physics; also, there are strategies to modify quantum mechanics and experiments that people have begun to perform.

ETHICAL ISSUES

Will awareness, or even perception of oneself to the general condition, be cultivated by an equal traditional chip, or are quantum impacts essential to have a feeling of "oneness"? Following Lawrence Krauss, "You should be alert if you hear something like, 'Quantum mechanics accompanies you with the world'... or then again 'quantum mechanics connects you with all fixings' It is conceivable to begin to be incredulous that the speaker is attempting to use quantum mechanics to generally assert that you could change the world by thinking about it." An abstract inclination is not sufficient to create this assurance. Individuals do not incline as numerous capacities do. Per Daniel Dennett, "As far as this subject is concerned, everyone is an expert... yet, they accept to have a particular individual authority regarding the quintessence of their cognitive encounters that can override any assumptions they find unsatisfactory." Doing tests to show quantum impacts requires experimentation on the human psyche since individuals are the main animals that can convey their experiences.

QUANTUM PHYSICS

CHAPTER 21

The Quantum Dimensions

More than two hundred years ago, quantum mechanics had its genesis in a simple experiment devised by Thomas Young, a British researcher in many sciences and humanities fields. The investigation requires only a light source, a table with two slits, and a screen on the other side to capture the light passing through the slits:

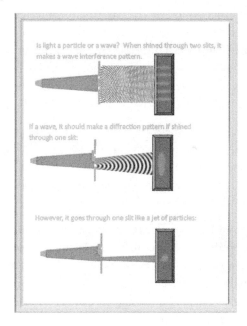

When Thomas Young reported the first "double-slit" experiment in 1801, the scientist who became Britain's Lord High Chancellor decried it as "devoid of any merit" and "the unmanly and fruitless pleasure of a childish and prurient fancy."

What did Mr. Young do to provoke such resentment from a country known for its culture of understatement? He showed that light manifested a dual particle/wave nature. But it was not the particle/wave nature of classical physics that he thought it was. It turned out to be the root of quantum mechanics. The experiment remains as unexplained today as it was then.

Young showed that when light passes through two slits, it resembles the familiar pattern waves make when an object is splashed in water or makes a noise in the air. Water and sound waves are propagating waves of standing water and air molecules that transmit energy by colliding. Waves of this nature interfere with each other when they are emitted from two sources. At some points, the waves manifest constructive interference where their crests combine to make more prominent crests and troughs combine to make deeper troughs. At other issues, there is destructive interference where the ridges and furrows cancel each other out. Noise-cancelling devices work by emitting "anti-noise" signals out of phase with the ambient noise so that the sound waves cancel out through destructive interference. Light makes the same constructive and destructive interference patterns when it passes through two slits, thus reinforcing the theory that it travels as a wave.

Then Young blocked one of the slits, expecting the wave behaviour to continue, as shown in the center of the image. However, this did not happen. Instead, the light shot through the single aperture like a jet of water moving through the air; it behaves like a wave when it passes through two slits and like a jet of particles when it passes through one.

Young's idea that light behaves like a wave was called "pandering" because Isaac Newton's scientific legacy was influential, and Newton had theorized that light travelled as particles. But perhaps light has a dual nature. Perhaps it travels as particles through space that generates electromagnetic fields as they move. Possibly when light passes through two slits, the electromagnetic fields interfere with each other like water waves. Still, when light passes through a single slit, the fields do not interfere, and the photons travel through space as particles.

Eventually, it was shown that light moves through space just like photons generate oscillating waves of electric and magnetic fields as they move. This made light behave like a "classical" wave of water or sound that interferes with itself as it passes through slits. This type of interference is known as diffraction. However, the slits must be microscopic for the tiny electromagnetic waves to bend around objects and cause diffraction interference. This effect is not visible with large objects. For example, if you put a physical barrier between you and the sun, you don't see the light bending in the shadows. However, the noise of a passing bird "chirping"

QUANTUM PHYSICS

behind the barrier is not blocked because the noise is a kinetic wave through a medium of air that bends around surfaces. The light interference Young saw through the double slits was a different phenomenon. Assuming it was caused by "waves" turning around the edges of the slits would be as fallacious as assuming that all water waves caused by wind, tides, and tsunamis have the exact origin.

In 1983, it became possible to fire photons through the slits one at a time. But, unfortunately, photons one at a time also generated interference. How could a photon interfere with itself? This could only happen if photons travelled as diffuse "waves" much larger than the electromagnetic waves we already knew. But if each photon travels as an extensive diffuse "classical" wave, we would expect most of it to hit the opaque barrier around the slits. At the same time, a minor part of it would pass through to impact the measurement screen to make an interference pattern from small pieces of each photon.

However, when photons are shot through the double slits one at a time, it is an all-or-nothing event. Photons are always detected as whole units arriving at a point. Electronic devices and our eyes see them that way. Either the entire photon goes through the slits and lands in one piece on the other side, or none arrives. The barrier stops most photons. Those that pass through the slits build interference patterns one by one on the measurement screen, as they do when trillions of slits pass through the slit simultaneously in a beam of light.

A photon interfering with itself is therefore inexplicable from classical wave mechanics. However, photons are massless particles that travel at the constant speed of light without experiencing the passage of time. Perhaps this is what allows a photon to interfere with itself. Then it was shown that the same thing happens with electrons, atoms, and molecules made up of many atoms. These are particles with mass that travel at less than the speed of light and therefore experience the passage of time, so it is not even theoretically possible for the same particle to interfere with its past or future incarnations. Like photons, these mass particles create interference patterns when passing through two slits while creating jets of particles when passing through one. It seems that all objects do this, up to some arbitrary and yet to be the determined size. Thus, these mysterious waves seem to apply to everything. It must be that the particles move through space as a diffuse "cloud" over a large area. When a particle cloud encounters a barrier with two slits, either it materializes as an impact point on the border, or it passes through both slits as two clouds that spread beyond the wall and interfere with each other, thus deforming the impact point of each particle into a pattern of interference bands that occur after many particles are shot through the slits one by one:

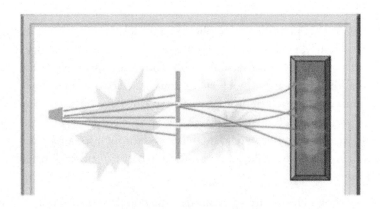

When a particle cloud encounters a barrier with a single open slit, the particle either materializes as an impact point on the border or passes through the slit in a straight line without interference and materializes as an impact point on the measurement screen. When many particles are shot through the single slit one by one, they land close together, creating a "clump pattern."

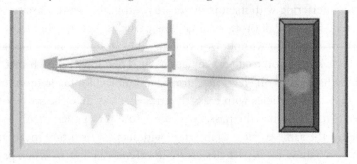

It gets more intriguing. We don't need to block one slit to eliminate the interference pattern and land the particles in a cluster. Instead, we can leave both slits open and stop the interference pattern by identifying which slits each particle crosses. Of course, we could do this by placing a detector on each slit, but it turns out that we only need one sensor placed on one of the slits.

If the detector is turned off, as shown in the top half of the image below, the particles create an interference pattern. However, as soon as the sensor is turned on, as shown in the lower half, the interference pattern changes to a clustered design:

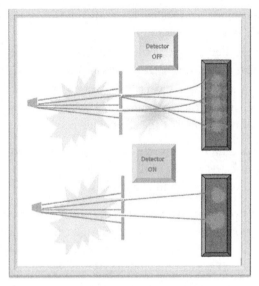

If the electron passes through the slit with the detector, its electric charge imparts information about its position to the detector, which reveals the electron's position and causes it to materialize in reality as a straight-line particle track that does not interfere with itself. An interaction has occurred whereby the electron's electric field has affected the electrons in the detector, causing it to send a signal to a memory device that records the electron's passage. We could theorize that the detector has also influenced the electron in some way that causes the electron to materialize.

If the electron passes through the slit that does not have the detector, then there is no interaction. Yet even the electron materializes in a particle that does not interfere with itself. This is because we can infer that if the electron arrived at the measurement screen without activating the detector, it had to pass through the slit without the detector.

Thus, it appears that this is not an interaction. Again, rather an even inferred information of the path a particle takes through the slits is what converts it from a cloud into a particle without any interaction with a detector. Once it becomes a particle, it takes on defined characteristics.

This ability to manipulate a particle by controlling the information we know about it implies that we may have the power to control some aspects of the Universe with our minds, controlling how much information we choose to observe about it. Of course, these tiny particles are so small as to make no difference in and of themselves. Yet, some interpretations of quantum mechanics postulate a quantum chain reaction from the smallest particles to the largest, all the way to the Universe itself. Since a photon or electron is part of the Universe, changing the state of our knowledge of these small particles affects our state of knowledge of the entire Universe and thus changes it.

This notion of changing the Universe by getting information about a subatomic particle seems absurd until we think about how we can start wonderful chain reactions in hydrogen bombs.

A few quintals of hydrogen are enough to destroy a large metropolitan area after igniting a fusion chain reaction in a hydrogen atom. We could, with modest effort, chain together enough hydrogen atoms to blast the surface of the Earth into space. If we can manipulate something tiny by controlling how much information we know about it, there may be no upward limit to the chain reaction we can manipulate. Since our minds store information, we ask: do they have the power to shape some aspect of the Universe?

We theorize that our brains might function on two levels:

Level I. Our brains are electromechanical "machines" that store information like the memory of a computer. If this is the highest level our brains are capable of, then consciousness is our mind's illusion to make us believe we are making decisions when we are not.

Level II. Our consciousness is a product of the mind, separate from the universe of matter, energy, space, and time. Perhaps our minds are so separate from the material Universe that they cannot be explained by material processes such as the electrical impulses in our brains that affect its atoms and molecules.

QUANTUM PHYSICS

CHAPTER 22

Examples and Applications

The theory, like classical mechanics, deals with the motion of particles in space and time. The difference lies only in the fact that classical mechanics describes deterministic continuous motion, while quantum mechanics describes particles' erratic, random movement.

Even though the new formulation of quantum mechanics makes the theory as understandable as classical mechanics, quantum phenomena are still strange to us who live in a classical world. In this part, we will give several examples to illustrate that the strangeness of the quantum world, which is missing in the everyday world, has all originated from the new particle motion and its new laws, for example, the discontinuity and randomness of action, the superposition principle, the collapse of the wave function, etc. These examples can help people understand quantum mechanics more deeply.

Schrödinger's Cat

In 1935, inspired by the famous paper written by Einstein, Podolsky, and Rosen, Schrödinger proposed the most renowned Gedankenexperiment, called "Schrödinger's cat paradox." The experiment was described by Schrödinger as follows:

A feline is confined in a steel chamber, next to the insidious contraption that accompanies it (which must be secured against direct obstruction by the cat); in a Geiger counter, there is a pinch of a radioactive substance, so tiny that perhaps in an hour one of the iota rots, but moreover, with an equivalent probability, maybe none; if it occurs, the counter cylinder disengages and through a transfer delivers a hammer that breaks a small jug of hydrocyanic corrosive. If you left all of this structure to itself for 60 minutes, you could say that the feline, in spite of everything it lives assuming then, has not rotted. The main nuclear rot would have damaged it. The ψ capacity of the whole framework would communicate this by having the living feline and the dead feline (pardon the articulation) mixed or scattered in equivalent parts.

So, when a decaying molecule cooperates with a feline utilizing a lot of gadgets, including a marker, a sled, and a small jug of hydrocyanic corrosive, and so on, the whole framework will be in a trapped superposition containing two branches as per the direct Schrödinger condition. In one chapter, the particle rots and intoxicates the feline, while in the other branch, the molecule does not decay, and the feline, nevertheless, lives. As a result, the kitten will be in an overlap of dead state and alive state. Let's first see what Schrödinger's cat looks like according to the image of discontinuous random motion of particles. The following figure represents Schrödinger's cat at six nearby instants. At each moment, the cat is either alive or dead in a purely random fashion.

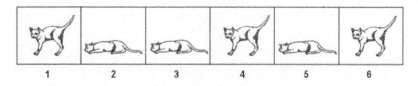

Figure1.3 Instantaneous images of Schrödinger's cat

However, during an interval of time, the random discontinuous motion of the cat will form a cat cloud, which is quite similar to an electron cloud. In one branch, the cat is alive, and in the other, the cat is dead.

The problem is how to reconcile Schrödinger's bizarre state of the cat with our macroscopic experience? This is essentially the measurement problem. The new formulation of quantum mechanics provides a straightforward solution, although some details are still missing. In Schrödinger's cat experiment, the atom is effectively in a superposition of the undecayed and decayed states. The state evolution is also governed by the linear Schrödinger equation very precisely since the effect of stochastic growth is minimal; for example, the collapse time of this superposition state can be longer than the age of our universe. However, when the atom interacts with the Geiger counter or cat and its superposition state is entangled with the state, stochastic evolution becomes significant. Thus, the entire superposition collapses almost immediately into one of its branches. As a result, the Geiger counter and the cat will always be solid-state; the Geiger counter either registers a particle or no particle, and the cat is either alive or dead.

QUANTUM ENTANGLEMENT AND NONLOCALITY

This part will present a clear physical picture of quantum entanglement and nonlocality, which are widely regarded as the most puzzling phenomena in the quantum world.

Let us first recall the single-particle picture. For the arbitrary intermittent motion of a molecule, the molecule tends to be in any conceivable situation at a given time. The probability thickness of the molecule occurring at any position x at a given time t is dictated by the square modulus of its wave work, specifically $\varrho(x, t) = |\psi(x, t)|2$. The physical picture of the molecule's motion is as follows. At the moment, the molecule randomly stays in one position. At the instant, it will either remain there or randomly appear at another work, probably not in the vicinity of the past position. Thus, during a time interval much larger than the duration of a discrete instant, the particle will move discontinuously throughout space with position density $\varrho(x, t)$.

Since the distance between the positions occupied by the particle at two nearby instants can be very large, the jumping process is nonlocal. In other words, two event instants of the particle (t1, x1) and (t2, x2) can easily satisfy the spatial separation condition $|x2-x1| > c|t2-t1|$.

Let us turn to the motion of two entangled particles. For the discontinuous random motion of two particles in an entangled state, the two particles have a joint propensity to be in two possible positions. The probability density that the two particles appear in each pair of classes x1 and x2 at a given instant t is determined by the square modulus of their wave function at that instant, i.e., $\varrho(x1, x2, t) = |\psi(x1, x2, t)|2$.

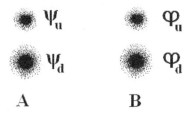

Figure 1.3 Two particles in an entangled state

Suppose two particles are in an entangled state $\psi u \phi u + \psi d \phi d$, where ψu and ψd are two spatially isolated conditions of molecule 1, ϕu and ϕd are two spatially separated conditions of molecule 2, and molecule one and molecule two are also in isolated areas A and B. The physical picture of this trapped state is as follows. Particles 1 and 2 are randomly in the $\psi u \phi u$ or $\psi d \phi d$ condition at the time, and afterward, they will also now remain in this state or jump to the next state at the time. Thus, during a short period, the two particles will spasmodically move through the $\psi u \phi u$ and $\psi d \phi d$ states with a similar probability of 1/2.

On this line, the two particles structure an indistinguishable whole and pass through the synchronous jump. In an emotional moment, if molecule one is in the ψu or ψd state, at that point molecule two must be in the ϕu or ϕd condition, and vice versa. Also, when molecule one jumps from ψu to ψd or from ψd to ψu, molecule two should all the time jump from $\varphi u to \varphi d$ or to $\varphi d to \varphi u$, and vice versa. Note that particle two is irrelevant to the distance between them and can only be explained by the joint propensity of the two particles as a whole. This gives us a clear physical picture of quantum entanglement.

Third, we analyse the nonlocal characteristic of the above-entangled superposition collapse during a measurement. The property of quantum nonlocality was brought to the attention of the physics community by Einstein, Podolsky, and Rosen in 1935. John Bell made a breakthrough in the study of quantum nonlocality in the 1960s. Bell's famous theorem shows the contradiction between locality and quantum mechanics (especially the collapse postulate). Many experiments have been conducted to test Bell's theorem to date. Although the results confirm the predictions of quantum mechanics and reveal the actual existence of quantum nonlocality, its nature is still a mystery. As Bell once said, "The scientific attitude is that correlations require explanation." The discontinuous random motion of particles and their laws can provide a natural explanation for nonlocal quantum correlations.

Suppose we measure the position of particle 1 in region A. Let the initial state of the measurement device be $\varphi 0$. When the measurement device interacts with particle 1 in a local area, the state of the measurement device will be entangled with the entangled state of particles 1 and 2, and the state of the combined system will become $\psi u \phi u \varphi u + \psi d \phi d \varphi d$. Since the stochastic evolution of the state will take effect during the measurement process, the state of the integrated system will soon collapse into one of the branches $\psi u \phi u \varphi u$ or $\psi d \phi d \varphi d$ with the same probability 1/2. As a result, the combined system as an entangled whole is disassembled into three independent parts:

the measurement device, particle 1, and particle 2. At the same time, the original two-particle whole no longer exists, and its state collapses to ψuφu or ψdφd, i.e., the state of particle 1 collapses to ψu or ψd, and the state of particle 2 collapses to ψu or ψd. Thus, the local measurement device carries a nonlocal influence on the two-particle system and especially on particle 2. Unification of two worlds

We live in a classical world. As a result, the objects around us seem to move continuously. In the quantum world, however, every particle moves in a purely random and discontinuous way. If the motion of all objects is essentially intermittent and unexpected, then why does the movement of macroscopic objects appear continuous? In this part, we will briefly explain how the transition from the quantum to the classical world takes place and why the discontinuous random motion of particles can provide a uniform picture for both the microscopic and macroscopic worlds.

We will first state the laws of random discontinuous motion more explicitly. Then, although the complete rules of action are still unknown, we can formulate its general form. According to our past analysis, the (non-relativistic) evolution of the wave function will be governed by a revised Schrödinger equation that contains two types of evolution terms. The first is the deterministic linear Schrödinger evolution term, and the second is the nonlinear stochastic evolution term that results in the dynamical collapse of the wavefunction. The equation can be written formally in the discrete form:

$$\psi(x, t+T_p) - \psi(x,t) = \frac{1}{i\hbar} H\psi(x,t)T_p + S\psi(x,t).$$

Where Tp is Planck time, the duration of a discrete instant, H is the Hamiltonian of the system for linear Schrödinger evolution, and S is the stochastic evolution operator for stochastic nonlinear development. Note that all quantities in the evolution equation must be defined in discrete space-time.

In the complete evolution equation, the linear Schrödinger evolution term will lead to the wavefunction diffusion. Conversely, the nonlinear stochastic evolution term will lead to the collapse or localization of the wave function. If the system's energy is minimal, then the evolution will be dominated by the diffusion process. This is exactly what happens in the microscopic world; a particle can pass through two slits simultaneously in the double-slit experiment.

y
CHAPTER 23

Test Realization of the Quantum Computer

The engineering simplicity makes the quantum computer faster, smaller, and less costly, be that as it may, its reasonable complexities are presenting troublesome problems for its test recognition. Various efforts have been made towards this path resulting in 20 enhancements. It is anticipated that time will not be too far off when the quantum computer supplants an advanced computer with all its possibilities.

Heteropolymers

Teich designed and worked on the first quantum computer based on heteropolymers was designed and worked on in 1988 by Teich and later improved by Lloyd in 1993. In a heteropolymer computer, a direct exhibit of particles is utilized as memory cells. First, data are put away on a cell by transferring the relative particle into an excited state. Then, the guidelines are transmitted to the heteropolymer by laser beats of appropriately tuned frequencies. The idea of the calculation performed on the chosen iota is controlled by the shape and range of the moment.

Ion Traps

A particle trap quantum computer was first proposed by Cirac and Zoller in 1995, first performed by Monroe and associates in 1995, and then by Schwarzchild in 1996. The particle trap computer encodes information in particle viability conditions and vibrational modes between particles.

Theoretically, each particle is processed by a different laser. However, fundamental analysis has shown that Fourier changes can be evaluated with the particle trap computer. This, thus, prompts Shor to consider computation, which is based on Fourier changes.

Quantum electrodynamics cavity

The quantum electrodynamics (QED) vacuum PC appeared by Turchette and associates in 1995. The PC includes a QED well loaded with cesium particles and a game plan of lasers, phase motion identifiers, polarizers, and mirrors. The goal is a genuine quantum PC since it can make, control, furthermore, ensure superposition and traps.

Nuclear Magnetic Resonance:

A nuclear magnetic resonance (NMR) PC includes a case stacked with liquid and an NMR machine. Each iota in the liquid is a free quantum memory register. They are counting proceeds by sending radio pulses to the test and scrutinizing its response. Qubits are realized as spin states of particle centres involving iotas. In an NMR PC, the memory register readout is practiced by an evaluation performed on a state factual suit, 2.7x10 19 particles. This is instead of the QED pit PC molecule trap, in which a separate singular quantum system was used for the memory register.

NMR PC can illuminate NP (Non-polynomial) complete questions in polynomial time. Most of the practical results in quantum computing so far have been cultivated using NMR PC.

Quantum dots

Quantum PCs based on the quantum dot innovation use a more straightforward design and less advanced testing, theoretical and mathematical capabilities when appearing differently than the four quantum PCs referred to recently. An assortment of quantum bits, in which touches are correlated to their nearest neighbours by techniques for gated tunnelling limits, are utilized to make quantum portals using a split input method. This plans one of the central points: the qubits are electrically controlled. The weakness of this design is that quantum touches can talk to their nearest neighbours. Therefore, simply reading the data is problematic.

Josephson Junctions

The Josephson cross-point quantum PC was demonstrated in 1999 by Nakamura. In this PC, a Cooper pair box, a touch of anode superconducting island, is weakly coupled to a bulk superconductor. The weak coupling between the superconductors makes a Josephson convergence, which carries on as a capacitor. If the Cooper box is as little as a quantum bit, the charge current breaks into the discrete motion of the single Cooper set. Finally, a single Cooper pair can be moved to the right convergence. Like quantum touch, in Josephson's convergence PCs, the qubits are electrically controlled. Josephson's convergence point quantum PCs are one of the promising opportunities for future advances.

The Kane Computer

This PC appears to be like a quantum dot PC, but indifferent habits are more like an NMR PC. It comprises a unique p 31 centre in an isotopically spotless, seductively inert Si 28 bead. The model is then placed in a powerful and attractive field to set the turn of p 31 equal to or antiparallel to the direction of the area. The lap of the p31core would then be constrained by applying a radio repetition heartbeat to a control anode, called an input, near the centre.

The interceding correspondence of electrons between turns could in this way be constrained by applying a voltage to terminals called J-doors, set between the p 31cores.

Topological Quantum Computer

The idea behind the topological quantum PC is to use the packet properties of braids representing the development of anyons to finish quantum computations. It is stated that such a PC should be impervious to quantum missteps of the topological quality of anyons.

CHAPTER 24

The Quantum World of Rotating Particles

Angular Momentum in Classical Mechanics

Before we focus on the mysteries and paradoxes of the quantum world, we need to learn a couple of elementary concepts - like angular momentum and spin - that we will find in the rest of this reading.

So, what is angular momentum? In CP, it's not very difficult to understand. It is rotational momentum, the momentum of rotating bodies. So, let's first look at the notion of angular momentum of a single particle rotating around a centre, as shown diagrammatically, with the vectors in Fig. 2.1.

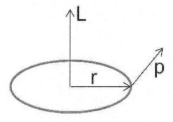

Fig. 2.1 The angular momentum of a particle.

Recall how the linear momentum of a particle moving unperturbed along a straight path is proportional to the product of its mass and velocity (at least in non-relativistic CM): = ·. Similarly, the general expression for the angular momentum of a particle with a rotational velocity around an origin is defined as:

= × = × ·, Eq. 1.

The position vector of the particle is relative to the rotational origin, and the cross denotes the vector or cross product (a kind of multiplication by vectors). Note that both linear and angular momentum are vector quantities, which means they have a magnitude and a direction. The letters representing the vectors are written with an arrow or in bold.

(The angular momentum vector L is perpendicular to the plane r x p. Its magnitude (length) is a measure of the speed of rotation. In the case of a body composed of n particles rotating about a centre external to itself (i.e., orbiting a centre of origin), the "orbital angular momentum,

can be obtained by determining the sum of the linear moments of each of its constituent particles-the aggregate over all the mass particles for the respective rotational velocity of each-as:

= Σ ×, again with I being the particle's distance from that axis and N being the number of particles. However, an extended body can also rotate on itself. A typical example is that of the Earth, which not only orbits the Sun but (in case you haven't noticed) rotates once around its polar axis about every 24 hours.

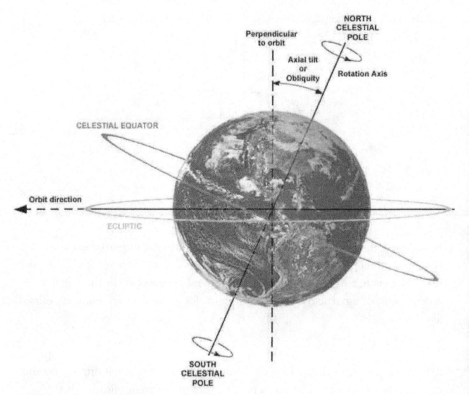

Fig. 2.2 The rotation of the Earth.

QUANTUM PHYSICS

In this case, for bodies whose interior has a complex mass distribution, such as the Earth (the density distribution within the Earth can be pretty complicated), one must calculate the so-called 'moment of inertia of a body.

Inertia is a physical quantity that measures the resistance a material body exerts to changing motion. In the case of a point particle, it is simply its mass. In the case of a body with a complex, more or less irregular extension and geometry and internal distribution of matter, it is a scalar quantity, an I-number, which mathematical procedures can calculate.

In general, the "angular momentum of rotation" is a vector given by:

$= \cdot$ Eq. 2

Where is the mathematical expression of 'angular velocity which is the measure of how fast it rotates around itself - that is, how quickly it completes a 360° rotation per unit time, which, in the case of a spherical body, is just =, the rotational velocity of the body divided by its radius.

Therefore, we must distinguish between the angular momentum of a particle or body moving around a centre (e.g., a planet around the Sun) and the angular momentum of a body around its central axis (e.g., the case mentioned above of the angular velocity of the Earth rotating around its polar axis). These are two closely related but slightly different quantities.

One is an orbital angular momentum, while the other is a rotational angular momentum, known simply as "spin." The total angular momentum is the sum of the two: $= +$.

An important universal law to keep in mind is 'conservation of angular momentum, which is a direct consequence of energy conservation (i.e., energy can never be created or destroyed, only transformed).

In Eq. 1, the radius or velocity can change, but the overall momentum L must remain constant. Thus, if a particle approaches the centre of rotation (r decreases), the angular velocity must increase to stay consistent. This is also true for any extended body. A typical example that illustrates this principle is that of the ice skater. When ice skaters bring their arms closer to their body, the angular velocity increases, and vice versa. This act, for example, of the ice skater, decreases the inertia I; according to the law of conservation of angular momentum, an increase must follow to keep it constant (see Eq. 2), and vice versa.

Fig. 2.3 Conservation of angular momentum

Now that we have been introduced to the main concept and principle of angular momentum in classical mechanics, let us apply this to quantum mechanics.

SPIN, THE STERN-GERLACH EXPERIMENT AND COMMUTATION RELATIONS

In QM, one is typically more interested in single particles or systems of many particles than in extended bodies. So, what about elementary particles, such as electrons? It is not difficult to extend the notion of orbital angular momentum of a material particle that has mass, such as, say, the electron in the case where we conceived of it in Bohr's atomic model, as a particle flying around the nucleus of the atom. This conception is quite doubtful because we should have in mind the atomic orbital not the Bohr atomic model. Anyway, so far, this picture of reality still works: The orbital angular momentum of the electron can be defined as the product of its mass times the velocity with which it orbits the nucleus times the orbital distance from it (Eq. 1 - this was also Bohr's mathematical approach that led to his atomic model). However, this is where the analogies to the classical world end.

In fact, what about the spin - i.e. what in QM is called "intrinsic angular momentum" - of a particle rotating around an axis passing through its centre, as in the case of the earth? The problem is that this analogy breaks down because we think of so-called 'elementary' particles as a point-like mathematical object. Thus, it becomes unclear what the spin of a point structure is supposed to mean. What is a point that spins on itself?

Intuitively and even mathematically, it doesn't make much sense.

So either the electron is not point-like, but no experiment has so far revealed any internal structure (at least not until 2020), or we have to view the notion of particle spin in QM as something that should not be viewed as we intuitively do at the macroscopic scale. For this reason, in QM, we speak more vaguely of particles "carrying" an intrinsic angular momentum, or spin, without literally suggesting the image of a spinning sphere, as in classical theory.

Yet, we know that even elementary particles like electrons, protons, and neutrons have a tiny but measurable spin. Can the spin of a particle as small as an electron be measured? The answer is positive and even surprising. It is possible to measure and associate spin to elementary particles. It is interesting that as we have seen that energy is quantized in QM, spin of particles is also quantized. Nature allows only discrete - not continuous - quantities of intrinsic angular momentum! Spin in QM is a quantized dynamical property of all particles.

How physicists arrived at this conclusion will be made clear with the Stern-Gerlach experiment. It is easier to understand how it works if we first introduce some basics and anticipate some results.

For elementary particles, spin has only two possible values. Thus, if we take the vertical axis as a convention (see Fig. 2.4) being the z-axis upward, we can show experimentally that, say, an electron has only two possible spins along that axis: a specific fixed amount of spin-up or spin-down states. We will never, ever observe the electron having any other amount of spin and being directed in another intermediate direction. Or, if you want to stick to the classical intuitive understanding, elementary particles rotate clockwise or counter clockwise around an axis, always with the same angular frequency.

To this picture we must add the fact that we must not forget that some particles, such as electrons and protons, are electrically charged particles. (Neutrons are not.) They all possess a small equal charge, which is also taken as the elementary electric charge, labelled 'e,' which amounts to

= 1.672 10 C, where [C] stands for 'Coulomb' and is the international standard unit

for electrical charge. This is an extremely small charge.

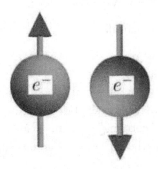

Fig. 2.4 A spin-up and spin-down electron.

For example, a typical household electrical device (e.g., a small lamp) operating with an electric current of 1A ([A] stands for 'Ampere,' the unit

of the intensity of an electric current), has something on the order of 6x10 (six quintillion, or six billion billion) electrons per second flowing through its circuit.

However, despite its small value, the electric charge of an electron can be measured by relatively simple experimental considerations and tricks. Electrically charged particles produce an electric field in their surrounding space, which interacts with other charged particles or magnetic fields. They themselves are also the source of magnetic fields.

Any electrically charged particle in motion always produces a corresponding magnetic field. A cable through which an electric current

- i.e. electrons flows, will also manifest a magnetic field. This is always true and is a fundamental law of physics: Wherever electric particles travel in space or through a conductor, they will always produce a magnetic field.

(The construction of electromagnets is based on this principle.) An electric current loop, a magnetic bar, an electron, a molecule, and a planet all have magnetic moments since they all contain circulating electric charges in one form or another. This is something we imagine happens with the Earth's magnetic field as well. The interior of the Earth must contain an enormous amount of magma, hot currents of a magmatic fluid, which is electrically charged. Its movement around the Earth's axis causes the Earth's magnetic field to build up. The opposite is also true. Wherever a magnetic field changes in time, an electric field will appear. This is what we have already mentioned when we discussed the nature of light as an oscillating EM field, and it is the basis of Maxwell's equations.

Therefore, an electrically charged rotating object is also expected to exhibit some magnetic field, since it is a rotating charge. The flow of charge around itself induces a magnetic field of a particle due to

its spin. This makes every elementary particle, such as electrons and protons, also small magnets. Therefore, not only do they possess an electric charge, but because they possess a spin, they must also produce a small magnetic field.

And since only two fixed spin states are possible, the result is a "magnetic moment", with the magnetic field lines directed in one or the other direction according to the spin orientation of the particle, as shown in Fig. 2.5.

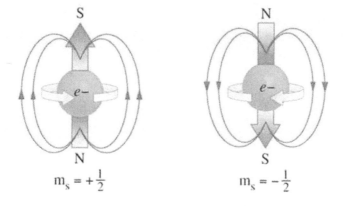

Fig. 2.5 Spin orientation and magnetic field lines of an electron.

CHAPTER 25

Five Modern Applications of Quantum Physics

The concept of Quantum Physics is difficult and strange. Understanding the activities of tiny particles and trying to define the forces that make them work brought Albert Einstein and his colleagues into a discussion on the subject. The problem with quantum physics is that it has a strange concept that defies common sense notions about causality, locality, and realism. Realism lets us know that something exists. For example, we might know that the sun exists even without looking at the sun.

Causality explains that something happens because something caused it to happen. If we flip the switch on a light bulb and see the light, that is causality. Because of the speed of light, when we light a match, it doesn't take a million light years for the light to come on; this is all location dependent. All these principles are not followed in the quantum realm. There it is, a whole other world.

A clear example of this is quantum entanglement, and this states that particles on opposite sides of the universe can be entangled to exchange knowledge instantaneously. This was a concept that Einstein could not accept. In 1964, a physicist named John Stewart Bell proved that quantum physics was a complete and viable theory.

Bell's Theorem

Bell's theory proposed a set of inequalities, now called Bell's inequality; this set represented how the spin measurements of an A particle and a B particle would be distributed if they were not entangled. As the experiment was being conducted, and after it had been

carried out, it was discovered that Bell's inequality was violated. He was able to show that quantum properties like entanglement are as accurate as fixing a tree.

And in this time, the various strange concepts of quantum physics have been applied to develop multiple systems with real-world applications.

Here are 5 of the most exciting ones:

1. Ultra-precise clocks

The need to have a reliable clock with precision is essential. Time is already synchronizing the technological world; time helps maintain stock markets as well as GPS systems. The standard watches we know make use of frequent oscillations of physical artifacts such as pendulums or quartz crystals to create their "ticks" and "touches."

In today's world, the most accurate clocks use the theories of quantum mechanics to calculate time. They control the specific frequency of radiation, which is needed to make electrons jump between energy levels.

The quantum logic clock, located at the U.S. National Institute of Standards and Technology (NIST) in Colorado, gains or loses only one second every 3.7 billion years. NIST's strontium clock, which was revealed not long ago, will remain accurate for a period of five billion years, a time longer than the current age of the Earth.

The importance of this super-accurate and sensitive clock crosses several industries: telecommunications, GPS navigation, and sensing. The accuracy of these atomic clocks depends in part on the number of atoms used.

When scientists try to cram a hundred more atoms into an atomic clock, it will make the clock's accuracy about ten times higher.

2. non-decipherable codes

In the traditional form of cryptography, keys were used to make it work. The sender of the secret message uses a particular type of key to encrypt the information, and the recipient may use another key to decrypt the information or news. However, the risk that this message may still be picked, cannot be taken for granted. To solve this problem, technologists have employed a theoretically unbreakable quantum key distribution (QKD). In QKD, polarized photons made to send the preliminary information are used.

This limits the photon; this only makes it vibrate in a singular plane. This could be up or down or left to right.

The other party accepting this data would then use the captured channels to decode the key and then use the chosen computation to interpret the message appropriately. Individual information is still sent on typical matching media, but one can only distinguish the news if it does not have the specific quantum key. The quantum decides that ecstatic "searching" photons will constantly change their states, and any push to listen will alert communicators about security penetration. On this day, organizations, for example, Toshiba, BBN Technologies, and ID Quantique, use QKD to plan super-secure systems. In 2007, Switzerland had the chance to embrace an ID Quantique item to give their races a carefully designed democratic framework. In 2004, Austria had the opportunity to receive trapped QKD to get the bank moving.

3. Superpowered Computers

PCs, for the most part, encode their data as a double string of digits orbits. Quantum PCs supercharge manipulation power because they use quantum bits or qubits that exist in a superposition of states - as long as they are not estimated, qubits can be both "1" and "0" simultaneously. Although the field is still being worked on, there remain hints of movement in the right way. D-Wave frameworks discovered their D-Wave One in 2011, which has a 128-qubit processor, and subsequently, D-Wave Two was revealed the following year and boasted a 512 qubit processor. A company report says these are the first commercially available quantum computers in the world. This has remained an issue because it's still unclear whether D-Wave's qubits are entangled. Research published in May finds signs of entanglement, but only in a reasonably small subset of the machine's qubits. There is confusion about whether the chips show actual quantum acceleration. There was a recent collaboration between NASA and Google to develop a D-Wave Two Quantum Artificial Intelligence Lab. Scientists at the University of Bristol were able to connect one of their traditional quantum chips to the Internet so that anyone with a web browser could learn quantum coding.

4. Enhanced microscopes

Research teams at Japan's Hokkaido University developed the world's first entanglement-enhanced microscope using the differential interference contrast microscopy technique. This particular type of microscope burns two beams of photons onto a material and tests the interference pattern changes depending on whether the beams touch a smooth or irregular surface.

Using entangled photons dramatically increases the amount of information the microscope can collect, as measuring an entangled photon provides information about its partner. Hokkaido constructed an etched "Q" that was just 17 nanometres above the surface with unprecedented clarity. Astronomical instruments such as interferometers can have their resolution increased using similar methods.

Interferometers are used to search for extrasolar planets, probe surrounding stars, and look for ripples in space-time or gravitational waves.

5. Biological compasses

The use of quantum mechanics is not something used only by humans. It is observed that birds like the European robin use this strange behaviour to keep track of their migration. They accomplish this process through a light-sensitive protein called cryptochrome, and this could have electrons tangled up in it. As soon as photons enter the eye, they reach the cryptochrome molecules. Enough energy is delivered to break them, and this forms two reactive molecules or radicals with the electrons split but still attached.

The lifetime of the cryptochrome depends mainly on the magnetic field around the bird. This is because birds have a susceptible retina, which can easily detect the presence of entangled radicals; this allows animals to notice a molecular-based magnetic map effectively.

Although, when entanglement becomes poor, the experiment showed that the bird will still detect it. Some types of lizards and insects also use this magnetic compass. Crustaceans, insects, and mammals. A particular kind of cryptochrome used for magnetic navigation in flies has been detected in human eyes. Whether humans have also used it for navigation is still debated.

Quantum physics as seen in everyday objects

At some point, we must feel annoyed and slightly confused about how so many concepts mentioned here can be applied to our daily lives or used by the different tools around us. Quantum physics is one of the highlights of human intellectuality, and its knowledge has helped shape our civilization. Despite this relevance, most people still feel that the subject matter of this field is quite abstruse and cannot be easily grasped by the ordinary mind. In the public's minds, quantum physics is seen as a problematic concept that is only understood by minds like Einstein and Hawking or another superhuman brain.

The concept of quantum physics is an understanding of the universe, and the universe is all around us, and its operation is based on quantum rules. Although we are so accustomed to the laws of classical physics, and this refers to the universe at the macroscopic level, the understanding of quantum physics still affects various familiar operations. Therefore, you will find different tools and equipment in this list that apply to the quantum principle without realizing it.

Toaster

We are all familiar with the red glow produced by the heating element as we toast bread. Interestingly, the observation of this red light led physicists to ask questions that gave rise to the quantum concept. Physicists wanted to know why hot objects glowed that particular red colour, a challenging question, and quantum physics came to shed light on this.

Max Planck answered this problem in his theory. He said that transmitted light must be discharged into discrete pieces of vitality, actual products of short, coherent occasions the recurrence of light. For high-frequency light, the quantum of energy is greater than the share of thermal energy, which is assigned to that frequency, making it impossible to emit light at that frequency. This prevents the emission of high-frequency light. Thus, we could say that the toaster could be a central place where the idea of quantum physics was born.

Fluorescent Lights

The traditional incandescent light bulb was able to emit light by properly heating a piece of wire until it became hot and cast a bright white glow; this is similar to the phenomenon of the toaster. You're enjoying a revolutionary work of quantum physics every time you turn on a fluorescent bulb or one of the newer CFL bulbs; this is quantum physics at work.

In the early 19th century, physicists discovered that all elements in the periodic table have a unique spectrum. When we heat a vapor of atoms, they will eventually emit light at a small number of discrete wavelengths, and each of the different elements will have a different pattern. Spectral lines have been used to classify the composition of new materials and unknown ingredients such as helium, for example. These materials/elements were first discovered through this process.

Here's how a fluorescent bulb works: whether the bulb is CFL or long tube, inside the bulb is a small amount of mercury vapor that is excited into plasma. Mercury efficiently emits light at frequencies that fall in the visible spectrum, and the eyes perceive it as white light.

QUANTUM PHYSICS

# CHAPTER 26

Modern Physics

This image, from CERN's Atlas detector, shows a Higgs boson decaying into two photons. Several recent discoveries affect how we understand how things work, from the human body to the universe!

Matter and antimatter

Figuring out what "matter" is made of is a mystery that has haunted philosophers since the beginning. Finally, in the 1930s, we had a partial answer, summarized in the table, where we have arranged the five particles we have discussed so far in approximate order of discovery.

Particle	Date	Mass (MeV)	Charge	B	Strong	EM	Weak	Grav
Electron	1897	0.511	-1	0		X	X	X
Photon	1900	0	0	0		X		X
Proton	1917	938	1	1	X	X	X	X
Neutron	1932	939	0	1	X	X	X	X
Neutrino	1933	< 10⁻⁶	0	0			X	X

The column titled "B" shows the number of baryons: we saw earlier that a neutron could not decay into two photons, and we explained this by introducing a new quantity that counts the number of nucleons present. Note how arbitrary the table seems: for example, we know that neutrino has a non-zero mass, but we do not know how big it is. Also, although it describes the particles, it does not explain them in any way.

Previously, we wrote the Dirac equation that describes the electron much better. Buried in this equation is one of the most remarkable discoveries in physics.

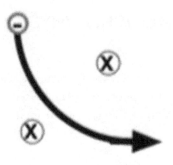

The Dirac equation has a solution for a negatively charged electron in a magnetic field, showing that it curves to the left for an area going off the page.

However, it also has a solution for an object identical to the electron, except it is positively charged, so it bends to the right in the same field.

In other words, Dirac's equation predicts antimatter! As Paul Dirac himself said:

"This result is too good to be false; it is more important to have beauty in one's equations than to make them fit the experiment."

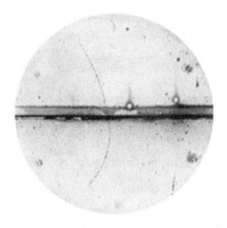

The prediction was confirmed very quickly by (in particular) Carl Anderson, who called the particle the positron. A positron is coming from the top of the image, passing through a skinny layer of lead, losing energy, and continuing with a lower momentum.

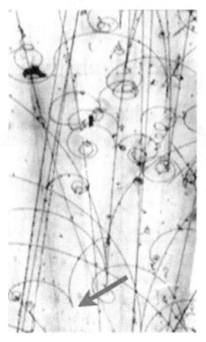

In this image, you can see many electron-positron pairs created in a hydrogen bubble chamber. The arrow points to a point where a couple has appeared out of nowhere: the process is

$\gamma \rightarrow e+ + e-$

Notice that we are creating matter from energy.

In solid matter, a positron will almost immediately encounter an electron and annihilate $e+ + e- \rightarrow \gamma + \gamma$. As a result, the photons emerge "back-to-back" with very characteristic energy given by half the total energy of the electron-positron pair, providing 511 keV per photon.

This is exploited in positron emission tomography (PET). Some isotopes decay with the emission of a positron (e.g., fluorine $18F \rightarrow 18O + e+ + \nu$), which immediately annihilates giving photons. These can be analysed to provide the site where the decay occurred. In the PET scan, the subject is given a form of glucose with an 18F atom replacing one of the hydrogen atoms in it.

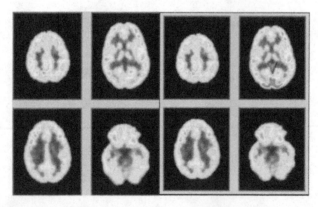

Any cell in the body actively using glucose will absorb it so that you can create a series of 2-D images. For example, above, you can see a brain at rest on the left, on the right where colours and patterns stimulate the brain: the stimulated area is the visual cortex, so you can see someone thinking! Interestingly, the most common use of antimatter is to understand the human body, but the radiation dose of 18F is shallow.

The electron is not unique: all particles have antiparticles to create antihydrogen, consisting of a positron bound to an antiproton. Antimatter has inspired many other ideas. For example, it has been suggested as an interstellar rocket fuel because 100% of the mass becomes energy when it annihilates, unlike a nuclear fusion reaction where maybe 2% could be used. The downside is that it is costly to produce and impossible to store with current technology: the best we can do is keep a few dozen atoms in a magnetic field.

In science fiction literature, the positron first appeared in Isaac Asimov's robot stories ("Caves of Steel," "I, Robot") as the core of the "positronic brain." Wisely, Asimov never attempted to explain how it might work. More recently, antimatter as an explosive is the theme of Dan Brown's "Angels and Demons."

Medical Imagin

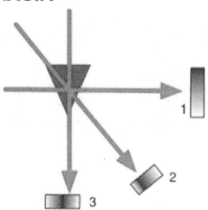

Principles of CT scanning

The medical industry needs to look inside the human body without damaging it. Physicists have provided a wide variety of techniques to do this, starting with Roentgen X-rays. However, a single X-ray image produces little more than a shadow-gram of the object, so the triangle image conveys little information. A series of photos give more but combining many pictures from different angles allows them to be incorporated into a 3-D image of the body. This is known as computed tomography or CT scanning.

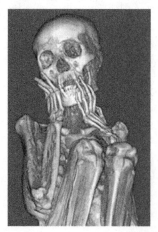

Peruvian Mummy

According to the non-invasive nature of it all, CT scanning can be used for delicate specimens, such as the ominous-looking Peruvian mummy in the figure.

EEG and ECG measure electrical signals from the brain and heart, respectively, while PET and various nuclear tracers allow us to see how metabolic processes work and how strong the bones are. However, the most powerful technique is MRI.

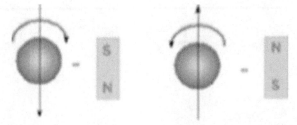

Because protons are charged and rotate, they behave like tiny magnets in the same way as electrons. Rabi first measured the nuclear magnetic moment in 1938. This led to the invention of magnetic resonance imaging in 1973. If you put a magnet in a magnetic field, its energy depends on its orientation.

QUANTUM PHYSICS

Because of quantum mechanics, there are only two possible spin orientations aligned along the magnetic field (up) or in the opposite direction (down).

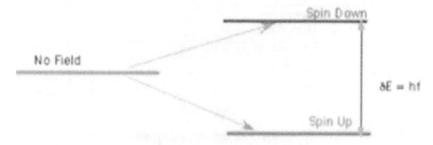

There is no magnetic energy in the absence of the field, so turning it on results in two possible energy levels divided by a minimal amount of energy. A common thing, the distinction between these vitality levels is comparable to a specific photon vitality/recurrence. For example, in the case the field B = 1Tesla = 10000 Gauss, at that point, the photon recurrence is f $\simeq$ 42.6 MHz, which is in the microwave district.

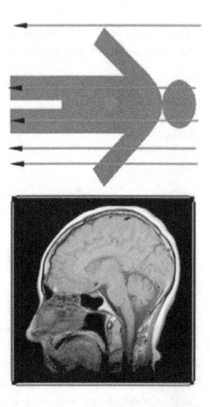

If protons are available, a sign taken up will be invested in this specific recurrence. Protons are present as hydrogen nuclei in the form of water and fat molecules but not in bone. To turn this into an image, we use a variable magnetic field, so the frequency of the emitted photon varies across the object. By scanning through a slightly different range of frequencies, we can build a 2-D image, which can then be assembled into a 3-D image.

In the diagram, you can see a "slice" through the head. Note that the skull bone appears white because it contains no water. MRI detects the difference between fat and blood vessels because of varying frequencies to produce a wonderfully detailed image. Unfortunately, this is the only method for detecting many tumours and showing "soft tissue" damage because X-rays cannot clarify the image's accuracy.

It is funny that the technique was initially known as Nuclear Magnetic Resonance but was changed to MRI to avoid the negative connotations of nuclear physics!

Some differences between the related techniques:

PET (positron emission tomography) uses

- X-rays

- Antimatter

- Proton magnetic fields

- Electrical signals from the brain

- Electrical signals from the brain

CT uses

- X-rays

- Antimatter

- Proton magnetic fields

Electrical signals from the brain

- Electrical signals from the brain

Magnetic resonance imaging (MRI) uses

- X-rays

- Antimatter

- Proton magnetic fields

- Electrical signals from the brain

- Electrical signals from the brain

- Elementary particles and the Higgs boson

QUANTUM PHYSICS

Is basic research worth it?

The effort to discover the Higgs boson is probably the most extensive scientific research effort in history. We have mentioned a few others, such as SNO. The cost is enormous: although discovering the Higgs boson was not CERN's only goal, the total cost of the LHC ranges from B$5 to B$8 approximately. However, it depends a lot on what is included in the price. How can we justify fundamental research when it has such a staggering associated cost and no apparent return?

However, it is essential to consider how powerful undirected research can be and how it can affect unrelated areas. To take one example, Rabi was certainly not trying to understand the human body when he discovered the proton's magnetic moment. Yet, MRI, based on his discovery, provides a unique way of imaging the body and is a billion-dollar industry today. In 1988, Tim Berners-Lee was a CERN employee when he invented the World Wide Web for geographically separated physicists to collaborate. Since CERN is an international organization, it has a policy of no copyrights or patents on inventions. The fortunes of all internet companies depend on Google, which alone is currently worth about $340B. In 2011 it had indexed 1 trillion pages, and a license of only .001 ¢/page would pay for CERN's annual budget!

In conclusion, Atlas and CMS give us the best control of our theories ever. However, good science costs money and requires effort. We don't know what it will produce (that's research!), and the payoffs may take 50 years to arrive.

Solutions

1. PET (Positron-Emission Topography) uses antimatter in the form of positrons.

2. CT scanning uses X-rays

3. MRI (Magnetic Resonance Imaging) uses the magnetic fields of protons.

CHAPTER 27

Modern Complication

Physicists can have difficulty when confronted with mysterious problems, not because they have more problems than if they were not impenetrable, but because popular culture tends to convolute the understanding of the mystery. It is no surprise that the media and movies can distort science when it does not specify precisely what it is talking about. It has been hypothesized that a different type of matter in our universe explains some of the mysterious observations. For example, galaxies are heavier than they should be, stars move faster than they should, and particles collide unexpectedly. Researchers in the field of astronomy have offered several explanations. In general, these explanations incorporate the existence of dark matter.

Dark matter is called "dark" because it is invisible to the human eye and any telescope technology built to date. This means, for research purposes, that dark matter cannot be easily tested, like particles that fall under the definition of "matter." For example, suppose you want to try Newton's laws of gravitation. In that case, you can take a banana, put it next to Venus, and measure the gravitational force on the banana as a function of whatever variable you want. You can look at the orbital acceleration of the banana, for example. Still, you can't find a dark banana, put it next to Venus, and figure out how it behaves in response to variable physical parameters. For example, it doesn't matter how ripe the banana is.

The universe is 27% dark matter, which raises a fascinating question: how is it possible that over a quarter of the universe is composed of something we can't even see? Also, how come there has been so much experimental support for something we can't work with?

A galaxy of dark matter?

The answer is that you don't need to see something to "do science" about it. Two branches of physics, in particular, have shown to be up-and-coming areas for experimentation: astrophysics and particle physics. For example, a team of researchers led by Peter van Dokkum at the W. M. Keck Observatory in Hawaii discovered an intriguing galaxy called Dragonfly 44. Dragonfly 44 has a small collection of stars. Van Dokkum's goal was straightforward: he knew that the stars would move at a certain orbital velocity. However, based on previous literature, he also realized that some stars in Dragonfly 44 were moving too fast for what should be their actual speed, according to the calculations. So the strategy was to estimate what the orbital velocity should be and measure the rates of the stars. Comparative analysis of these values might reveal interesting possibilities (or maybe not!).

The predicted orbital velocity of stars is not incredibly difficult to calculate. Two theories of gravitation can be used to do this: Newton's law of gravitation or Einstein's theory of general relativity. According to the former, the more matter there is in a system, the larger the magnitude of the force on each component. The greater the power, the greater the acceleration, and the more significant the change in orbital velocity. According to the second, gravity is related to the curvature of spacetime. Still, there is a similar argument: more matter in the spacetime map leads to more curvature, which means more substantial gravitational effects. Both situations postulate that the more concern there is in a given space area, the more significant the change in orbital velocity of the components (stars) that make up that system should be. Since there were relatively few stars in Dragonfly 44, it was assumed that they could not move very fast. However, when they calculated the experimental orbital velocities, they ended up with significantly higher values than they should have been. Newton or Einstein doesn't matter which side you're on because you'll eventually come to the same conclusion: if these stars are moving faster than they should be, there must be more mass in the system. Precisely, the group that scientists before Van Dokkum could not detect.

It fits the description of dark matter quite well. If Dragonfly 44 were composed of dark matter, it would have a mass that we cannot measure. This extra mass could contribute to the gravitational effects experienced by stars, which is why they move with such high orbital velocities. So, the next goal was to dig a little deeper: if we could calculate how much mass is needed to force stars to move at their speed, we could know how much dark matter is present in this galaxy. It turns out that. Dragonfly 44 is composed of nearly 99% dark matter. Thus, Van Dokkum and his team found a dark matter galaxy.

A particle accelerator at the CERN lab collides large hadrons or a specific class of particles. Despite nearly a decade of debate over a unique name for the collider, participating physicists agreed on the Large Hadron Collider (LHC). In the LHC, particles can be made to collide with each other to produce collisions. When two particles collide, two variables are important: the energy and the momentum changes of the crash. Generally, in ideally isolated systems, both the power and momentum of the collisions are conserved. For example, this means that if you send two bananas crashing into each other and watch the two bananas stick together after the crash, the energy of one banana added to the power of the other banana will give the total energy of the sticking bananas at the end. The same is valid for momentum.

Yet, this is not always the case. If baryons and fermions, members of other classes of particles, are sent in motion toward each other, conservation of momentum and energy is generally not observed. Energy and momentum both tend to be lost. However, if you force the baryons and fermions to collide ideally, that is, you deprive the system of everything but these two particles, you should achieve conservation of momentum. This is exactly what happened at the LHC: baryons and fermions were put into isolated systems with speed before and after the collision. What should not have been a difference became a pronounced difference: momentum was lost.

It was then postulated that dark matter particles were produced in these collisions. Such dark matter particles took away some momentum to create the change that was measured. Since dark matter can escape detection, the measurements needed to verify these postulations could not be made directly. However, the data were exceptionally telling when the momentum of these putative dark matter particles was measured and compared to the change in speed of these particle collisions.

Conclusion

erhaps, indeed surely, I will repeat myself a bit, but, as you know by now, this is one of my "strategies" of dissemination, which many appreciate and, therefore, I continue to do so.

First, let me make an observation that certainly is not new to you. Quantum Mechanics (QM) was born in the same years of the theory of relativity and has been, in a similar way, a reference theory for all the 20th centuries. However, it has never managed to get out of the narrow circle of insiders. You might think this is due to the mathematical difficulties of the expressions that govern the wave function and the complex plane and similar things. But, no, that is not enough to explain its "ghettoization."

There must be something else that seems to be blocking its revelation. Relativity is not like that, and yet it has forcefully entered the everyday language. Moreover, QM is at the premise of today's apparent multitude of mechanical developments, from nuclear vitality to PC microelectronics, from advanced tickers to lasers, semiconductor frameworks, photoelectric cells, and indicative and treatment hardware for certain diseases. To put it bluntly, we can "live" in a "cutting edge" way today, thanks to QM and its applications.

As I said before, our mind seems to be based on quantum processes, including state superpositions, wave collapses, and entanglement situations. The real difficulty lies in its "counterintuitive" postulates about the reality of nature—an actual discomfort in entering an unknown and absurd world like Alice's. Let's not feel too inferior, though. The founding fathers themselves experienced this on the edge of absurdity. Was it possible to believe that nature followed completely arbitrary rules or, instead, was it all an appearance due to the lack of information, of a deterministic type, still missing?

The same creator of the general and ultra-confirmed uncertainty principle (Heisenberg) said: "I remember the long discussions with Bohr, which made us stay up late at night and left us in a state of deep depression, not to say of true despair. I kept walking alone in the park, and I kept thinking that it was impossible that nature was as absurd as it appeared to us from the experiments." In short, there is no defined and describable reality, but an objectively indistinct reality composed of overlapping states.

Let us take up two essential points that we have come to know but certainly not to understand:

1. Every action of the most acceptable structure of matter is characterized solely and only by its probability of occurrence—completely causal phenomena, not deterministic. But, above all, by the indistinct separation between the observed object, the measuring instrument, and the observer.

2. It is possible that, under certain conditions, what happens in a particular place can drastically affect what happens in a completely different place instantaneously. This leads to the phenomenon of entanglement, the interweaving of particles that have had an interaction in their past (but recent research seems to admit also "contacts" in the future) or that were born "together." Although entirely separate, they always represent the same entity. Thus, an action taken on one has an instantaneous effect on the other.

Perhaps you have already noticed the fundamental problem of QM. On the one hand, the difficulty of dealing with concepts too far from everyday reality, and on the other hand, the test of using adequate language to explain this absurd world. Mathematics can also describe it, but the letters and words of this strange alphabet are missing.

Exceptional, in this sense, was the work of Feynman with his diagrams applied to QED (which we now know pretty well).

It is intriguing to quote a phrase by Max Born in this regard: "The definitive beginning of the problem lies in the reality (or philosophical rule) that we are forced to use the expressions of basic language when we want to represent a marvel, not through a legitimate or numerical examination, but through an image that speaks to the creative mind. Everyday language has evolved from everyday experience and can never overcome these limiting points. Traditional material science has limited itself to the use of such ideas; by examining perceptible movements, it has created two different ways of speaking to them by basic cycles: moving particles and waves. There is no other method of giving a pictorial representation of the motions - "we must apply it also in the area of nuclear cycles, where the old-style material science separates."

More or less, the representation of QM itself could be intensely influenced by our "artwork" graphical cut-off points.

Thus, the founding fathers themselves often used analogies and similes to express purely mathematical concepts. They must, however, be taken for what they are and not given any absolute, concrete validity. This is a huge problem for our brain (especially today) even if - maybe - it would have all the bases to use an adequate language, but still too indistinct to be formulated correctly: Feynman's diagrams, I repeat, are an excellent attempt in that direction.

Niels Bohr himself used graphic analogies to try to support theories so absurd for our classical language. Famous is the white vase that represents, at the same time, two black social profiles.

A state of superposition between two realities existing instantaneously (two states or - perhaps - two universes?). This type of analogy has influenced many optical illusion games and even artistic currents (think Picasso).

It is a pity that these interpretative efforts, together with the more complete and refined ones of Feynman, do not find their way into schools to adequately prepare young people to "stutter" their first quantum words and to start a primitive language that would allow them, today, to understand, at least partially, the reality of Alice. And not just passively undergo the most beautiful technological applications that are now an integral part of their physical bodies. Actual "appendages," which, however, act unconsciously, independently from any mental command. Unconditional reflexes and nothing more.

De Broglie advanced his bold hypothesis just following the symmetries of visible nature. He associated only to matter in general, what happened to light. In short: if the light is manifested under a double aspect, undulatory and corpuscular, why not think that matter follows the same rule? It is enough to associate a wave of a certain length to each corpuscle of matter, which is a phenomenon extended to the space surrounding the particle. The dualistic nature (particle-wave) applies to all particles, such as electrons, atoms, and other moving entities.

However, the fundamental problem remains (still subject of discussion and interpretation), which we have mentioned. The matter-wave that commands the particle can be deterministic, and therefore still unknown in its fundamental structure (in line with Einstein's idea) or, instead, a different representation of the same particle and thus follow the rules of a complete causality (Copenhagen school).

One way or another, however, it must be concluded that light or a beam of electrons is nothing more than a "train" of electromagnetic waves, but also a jet of "bullets" as in the double-slit experiment.

While remaining in this essential ambiguity, Schrödinger formulated the equation that perfectly describes every adulatory property of matter through its wave function. It allows one to explain every behavior and, more importantly, to calculate the probability of finding a particle within the associated wave. However, overwhelming mathematics does not negate the fact that Schrödinger himself did not believe in the actual concreteness of this representation. "Everything and the opposite of everything (conceptually), but described in the same way."

However, his equation confirms what Feynman's experiment illustrated above: a particle can occupy ALL possible positions within the associated wave. By occupying all possible situations, it no longer has an actual place of existence or direction. It automatically nullifies any possible prediction of its future except in purely probabilistic terms (QED is always more understandable... don't you think?). The pilot wave or a hidden variable does not change nature's action and its probabilistic description.

Once again, we fall back on Heisenberg's principle. In classical mechanics, the deterministic essence automatically allows one to predict the future if one has exact information about the position and velocity. We remember, in this regard, that the first mathematical methods that allowed the calculation of an orbit of a "planetary" particle were based (and are still based) on the knowledge of at least three positions and three speeds, such as to allow the solution of an orbit characterized by six unknowns. Too easy for microscopic particles.

The probabilistic conception inevitably leads to the uncertainty principle inherent to all microcosm: either you know the position or the speed. To understand both with precision is impossible. Otherwise, the particle would be localized, and the wave would collapse. And we go back again to the starting point. If there is an initial causality (completely unknown) or if there is no causality at all. In a nutshell, the double-slit experiment perfectly illustrates all the problems of QM.

It is worth reflecting on Einstein's dramatic emotional situation. While he was giving physical reality a perfectly deterministic representation, he found himself involved in a presentation that led to the complete causality of nature.

QUANTUM PHYSICS

He said," The theories of quantum radiation interest me greatly, but I would not want to be forced to abandon strict reason without trying to defend it to the limit.

Yet, no physicist has contributed as much as Einstein to the creation of quantum physics. What he demonstrated about it (and for it) was sufficient and advanced a scientific career of the highest order (not for nothing did it earn him the Nobel Prize). It is therefore easy to understand his existential drama, which never abandoned him until his death. A mixture of anger, wounded pride, unshakable confidence, and despair for not proving his certainties.

This mix of frustration, exaltation, hope, disappointment, innovation, and conservatism permeated all the great minds that gave birth to the QM. A very choral work and indeed not a puzzle of individual ideas. Almost unwillingly (sometimes even against their purposes), everyone did nothing but put one more brick to a building that was becoming an incredible skyscraper with increasingly solid and unassailable foundations.

Maybe this unique way in the history of science to formulate a more complete and refined theory, by many superior minds, could make us understand that QM is something inherent in the human mind but has extreme difficulty coming out. Indeed, the knowledge of the language of Classical Physics has made great strides, but not so far from the almost unconscious intuitions of Democritus and Epicurus. Simply put, the mind must be trained to go along with a reality that is only historically and culturally absurd.

The more we go into the essence of QM and its principles, the more the double-slit experiment becomes fundamental and complete. A proper scientific masterpiece, a manifesto itself of the future of human intellect.

QM not only as a science but as a school of life.

QUANTUM PHYSICS

Made in the USA
Las Vegas, NV
13 October 2023